HIGH ALTITUDE ENERGY

MINING THE AMERICAN WEST

High Altitude Energy: A History of Fossil Fuels in Colorado,
Lee Scamehorn

Yellowcake Towns: Uranium Mining Communities in the American West,
Michael A. Amundson

SERIES EDITORS

Duane A. Smith
Robert A. Trennert
Liping Zhu

HIGH ALTITUDE ENERGY
A HISTORY OF FOSSIL FUELS IN COLORADO

LEE SCAMEHORN

University Press of Colorado

Published by the University Press of Colorado
5589 Arapahoe Avenue, Suite 206C
Boulder, Colorado 80303

Printed in the United States of America

The University Press of Colorado is a cooperative publishing enterprise supported, in part, by Adams State College, Colorado State University, Fort Lewis College, Mesa State College, Metropolitan State College of Denver, University of Colorado, University of Northern Colorado, University of Southern Colorado, and Western State College of Colorado.

The paper used in this publication meets the minimum requirements of the American National Standard for Information Sciences—Permanence of Paper for Printed Library Materials. ANSI Z39.48-1992

Library of Congress Cataloging-in-Publication Data

Scamehorn, H. Lee (Howard Lee), 1926–
High altitude energy : a history of fossil fuels in Colorado / Lee Scamehorn.
p. cm. — (Mining the American West)
Includes bibliographical references and index.
ISBN 0-87081-661-6 (hardcover : alk. paper)
1. Fossil fuels—Colorado—History—19th century. 2. Fossil fuels—Colorado—History—20th century. I. Title. II. Series.
TP317.U5 S33 2001
333.8'2'09788—dc21

2001007760

Design by Daniel Pratt

11 10 09 08 07 06 05 04 03 02 10 9 8 7 6 5 4 3 2 1

CO-WINNER OF THE 2001 COLORADO ENDOWMENT FOR THE HUMANITIES PUBLICATION PRIZE

The CEH Publication Prize was created in 1998 and the first awards were made in 1999. The prize annually supports publication of outstanding nonfiction works that have strong humanities content and that make an area of humanities research more available to the Colorado public. The CEH Publication Prize funds are shared by the University Press of Colorado and the authors of the works being recognized.

The Colorado Endowment for the Humanities is a statewide, nonprofit organization dedicated to improving the quality of humanities education for all Coloradans.

Contents

Illustrations

Photographs

Figures

Maps

Tables

Preface

A nontechnical history of fossil fuels in Colorado, this book examines the origin, evolution, and some aspects of the social and economic impact of Colorado's coal, petroleum, natural gas, and oil-from-shale resources. The production and distribution of some of these abundant fuel resources became major industries.

The first was the production and sale of coal. The early producers were located mainly along the Eastern Slope of the Rocky Mountains, from Trinidad in the south to Erie and Marshall in the north. By 1900, coal had contributed, in one way or another, to virtually every aspect of human activity in the state.

Consumers of fuel, with some exceptions, used coal. It was burned in cooking ranges, water heaters, and stoves and furnaces in homes and businesses. In addition, it was the primary fuel for firing boilers to generate steam to drive the machinery of industry and commerce, including railroad locomotives. Furthermore, it was the raw material for making three important secondary forms of energy—coke, manufactured gas, and electricity.

All three played important roles in the evolution of the state's energy history. Coke, which was coal transformed in ovens to almost pure carbon, was consumed in the smelting of metallic ores. Manufactured gas, made by distillation of bituminous coal, was distributed in many towns and cities as fuel for lighting, cooking, and heating. It was eventually displaced by natural gas. Electricity, generated in coal-fired steam power plants, was consumed for lighting and to run appliances and machines in homes, offices, and industries.

In the twentieth century, coal gradually lost ground to oil and natural gas. Colorado's solid-fuel output exceeded 12 million tons in 1920, then gradually fell to less than one-fourth that amount in the 1950s. Severe contraction raised the specter of total collapse. That did not happen. In 1960 the production of coal rebounded to more than 3 million tons, and continued upward, slowly at first, then at a rapid pace in the 1970s. The turnaround was attributed to the rising demand for fuel by power plants. Coal-fired generating stations accounted for most of the state's electric power generation. At the same time, Colorado coal was shipped to other states where utilities burned low-sulfur coal in order to avoid the need for costly equipment to diminish sulfur dioxide and other toxic emissions.

Seepages near Cañon City were a source of oil in the 1860s, and production from deep wells began at nearby Florence in the 1880s. Local refineries produced lamp oil and lubricants, which the Continental Oil Company, an instrument of the Standard Oil monopoly, distributed throughout the Rocky Mountain region. Natural gas was usually a by-product of oil production until the 1920s, when commercial quantities were produced first in Larimer County, then in western Colorado. During that decade-long time, consumers of coal began using natural gas for space and water heating, as well as cooking.

In the twentieth century, petroleum was consumed in Colorado, as in the nation, primarily as motor fuel. The insatiable thirst of ever-increasing numbers of motor vehicles stimulated exploration and development of oil fields throughout Colorado. Production reached a high of 47.5 million barrels in 1960, then declined for the next four decades. A similar downward trend occurred in the forty-eight contiguous states. The shrinkage in domestic oil output, at a time when demand for petroleum products continued to rise, heightened the nation's dependence on oil from abroad.

When the flow of foreign oil was interrupted in the 1970s, western Colorado experienced an oil-shale boom for the third time in the twentieth century. The effort to extract oil commercially from the region's substantial shale resources crashed in the 1980s. Only one enterprise survived until the 1990s, when continuing low prices for crude petroleum prompted a decision to terminate technologically promising production methods. As long as the price of foreign oil remained below the cost of extracting oil from shale, or from coal, there was no incentive to produce synthetic oil.

The natural-gas industry, after struggling through decades of restrictive government regulations, expanded rapidly in the 1980s, when free-market policies were applied to interstate, as well as intrastate, transmission of the gaseous fossil fuel. Output ran ahead of demand in the early 1990s largely because of the rapid commercial development of methane from coal seams. Natural gas, the most environmentally benign of the fossil fuels, gained popularity in the 1990s, when public concern focused on pollution caused by the combustion of coal and petroleum.

For similar reasons, attention turned to alternative fuels, meaning nonfossil fuels. By the close of the twentieth century, renewable fuels in the form of solar, wind, biomass, and geothermal energy had achieved promising potential for further development. At this time it seems unlikely, however, that they will become the nation's principal source of energy in the foreseeable future.

High Altitude Energy is topically and chronologically organized. Chapter 1 examines the mining of coal and the use of that fuel from the time of the Pikes Peak gold rush to the close of the third decade of the next century. The extraction of coal was labor-intensive; towns to house miners and their families dotted the landscape of many counties, especially Boulder, Fremont, Huerfano, and Las Animas, east of the mountains, and eventually Delta, Garfield, Gunnison, La Plata, Pitkin, and Routt beyond the Continental Divide. Mines attracted immigrants whose cultures contributed to the cosmopolitan flavor of Colorado's society.

Chapter 2 looks at the manufacture of coke from coal during the years from 1877 to 1930. Coke was made by burning coal in ovens to eliminate impurities. The resulting high-carbon fuel was used for the reduction of precious metal ores in furnaces to produce gold, silver, copper, lead, and zinc. Beehive ovens were located at or near coal mines in southern and western Colorado, where there were abundant deposits of metallurgical, or coking-grade, bituminous coal. The men who worked the ovens, mostly immigrants, usually lived in the coal-mining towns. Coke was consumed in the blast furnaces of the Colorado Coal and Iron Company and its successor, the Colorado Fuel and Iron Company, the only fully integrated maker of iron and steel in the West until World War II. Blacksmiths burned coke in forges and, after the turn of the century, beet processors used it as a filtering material in making sugar.

Chapter 3 examines the evolution of Colorado's petroleum industry from 1860 to 1930. Initially, oil consumed in Colorado and in the Rocky Mountain region was from Cañon City and Florence, in the midst of Fremont County's coal camps. Until the turn of the century, crude oil was refined mainly to produce kerosene, which was consumed primarily as lamp oil. After 1900, the rising demand for gasoline as fuel for internal-combustion engines stimulated exploration and development of additional crude supplies. Boulder was the scene of the state's second oil field. Until 1930 Florence was the center of refining activity, producing the fuel and lubricants for a regional market.

Chapter 4 discusses the early development and utilization of manufactured and natural gas in Colorado. Without easy access to urban industrial markets, natural gas was little used and mostly wasted as a by-product of crude-oil production. Exploration and development of gas fields were delayed by the absence of pipelines capable of linking wells with the region's population centers. Natural-gas resources were not developed on a significant scale until the 1920s. However, gas was available to residents of most cities and larger towns. Urban-based utilities distilled coal in retorts, and distributed what was called manufactured gas through underground mains to residential and business customers for lighting and heating. Consumption for illumination declined after electric utilities provided power for homes, commerce, and industries. The state's manufactured-gas industry served customers with fuel for cooking and space and water heating until inexpensive natural gas was available.

Chapter 5 treats the production and consumption of coal from 1930 to the energy crisis of 1973. In that time the industry, after having served local and regional needs for half a century, entered into a long-term recession, before experiencing another boom in the 1960s. The temporary decline of coal consumption ended a way of life for many Americans. Householders, for example, abandoned coal for gas or oil as fuel for furnaces, water heaters, and cooking stoves. Railroads and other industrial consumers of coal adopted petroleum or natural gas. Only the production of metallurgical coal remained reasonably stable, as the West's newer integrated steelworks bought Colorado coal to make blast-furnace coke. In the 1960s, coal mining enjoyed a revival of prosperity because utilities built new coal-fired steam-generating stations in an effort to keep pace with the rising demand for electric power throughout the region. Technological advances, by lowering production costs, contributed to the solid fuel's resurgence.

Chapter 6 examines the evolution of the petroleum industry from the 1930s to the 1970s. During the 1920s and 1930s, consumption in the United States did not keep pace with the rapid exploration and development of petroleum resources. As supply outpaced demand, the price of crude oil plummeted until, by the 1930s, producers called for government intervention to prevent what they feared was the imminent collapse of their industry. States could not effectively respond to a problem that was national in character, forcing Congress and the president to initiate legislation designed to achieve fair and stable prices by balancing production and consumption. Successful in the short run, this policy became superfluous in the 1940s because World War II stimulated demand for, but not the supply of, petroleum products. By the time that conflict ended, a new era of prosperity challenged the oil industry to keep pace with skyrocketing consumption. Colorado's Western Slope counties were the principal source of crude oil until the 1950s, then the main thrust of oil and gas exploration shifted to the Denver Basin.

Chapter 7 examines the production and role of natural gas as a fuel for Colorado homes, businesses, and industries from 1930 to 1973. Until the 1950s, most of the state's gas fields and the principal urban markets were on opposite sides of the Continental Divide. Each was inaccessible to the other, forcing Denver and other eastern communities to secure natural gas via pipelines from other states. In the 1950s, the Denver Basin supplied the state's principal urban markets. Meanwhile, the growing volume of gas produced west of the mountains was largely marketed, via interstate pipelines, in the heavily populated Pacific Coast states.

Chapter 8 treats Colorado's abundant but largely neglected oil-shale resource. While shale, which was impregnated with a substance called kerogen, served as a source of oil for many countries, its commercial production in the United States ended soon after Colonel Edwin Drake drilled the first oil well near Titusville, Pennsylvania, in 1859. However, whenever a shortage of domestic petroleum supplies appeared imminent, private enterprise, with the encouragement of the federal government, sought to develop the oil potential of shale, particularly in the Colorado portion of the Green River Formation. Oil-shale booms were followed by busts because petroleum shortages were always of brief duration.

Chapter 9 is devoted to the energy crisis, a social and political reaction to an interruption in the flow of petroleum from the Middle East in 1973. This prompted a joint effort by the federal government and private enterprise to achieve national energy self-sufficiency. Project Independence sought to expand production of all fossil fuels, with special emphasis on the nation's abundant coal resources. Programs designed to expand the output of oil and natural gas, and to produce synthetic fuels, were coupled with nationwide programs to encourage the conservation of energy.

In Colorado, as elsewhere, the pursuit of energy independence faltered in the early 1980s, as cheap oil from abroad flooded the American market. This caused many people to question the cost of massive government programs designed to produce all of the nation's energy at home. Foreign oil undermined the short-lived revival of the domestic petroleum industry and ended corporate efforts to extract oil from shale and to develop new technologies for extracting oil and pipeline-quality gas from coal. The collapse of petroleum prices in the mid-1980s marked the end of the energy crisis.

The final chapter traces, in a general way, fossil-fuel production and consumption from the end of the energy crisis to the close of the twentieth century. While the output of coal continued to move upward, the production of petroleum fell because of declining reserves. Natural gas, once freed from restrictive regulations, expanded at an impressive pace. It was the fuel of preference for many consumers because it did not release significant amounts of harmful pollutants into the atmosphere to produce acid rain. However, reserves may not permit long-term, sustained development and utilization.

The consumption of coal and oil raised widespread concern about the risks their combustion posed for the public's health and the environment. The Clear Air Amendments of 1990 diminished the threat of acid rain, but did not alter what environmentalists called the problem of global warming, allegedly caused by a growing imbalance of greenhouse gases in the atmosphere.

Acknowledgments

The research for this study began in the 1970s as an attempt to understand what was popularly known as the energy crisis. The nation's brief commitment to energy independence, triggered by the interruption of oil shipments to the United States by the Arab states of the Middle East, had an enormous impact on Colorado. This is a history of fossil fuels, and their role in shaping the history of one state.

The author visited libraries in the Rocky Mountain region to collect data about the development and utilization of coal, petroleum, natural gas, and oil from shale in Colorado. The professionals who staff those institutions deserve acknowledgment of their valuable assistance. Particularly helpful were the dedicated people at the American Heritage Center, University of Wyoming, Laramie; the Archives and Manuscripts Department, the Harold B. Lee Library, Brigham Young University, Provo, Utah; the Western History Room, Pueblo Regional Library; the Arthur Lakes Library, Colorado School of Mines, Golden; the Government Documents Library and the Western Historical Collections of Norlin Library, the University of Colorado at Boulder; the Stephen H. Hart Library of the Colorado Historical Society, Denver; the Western History Department of the Denver Public Library, Denver; and the Boulder Public Library.

Doug Carriger read an early version of this study. Rosslyn Scamehorn performed, with her usual dedication and efficiency, the roles of research assistant, editor, and critic. The author is, of course, solely responsible for errors, whether of facts or interpretations.

HIGH ALTITUDE ENERGY

1

King Coal 1860-1930

At the time the Pikes Peak gold rush attracted fortune seekers to what became the territory of Colorado, most of the energy utilized in the United States in homes, commerce, and industry was derived from wood and coal. By the 1890s, coal had largely supplanted wood, accounting for more than 90 percent of the nation's consumption of energy; most of the remainder was attributed principally to petroleum and natural gas. An abundance of coal, and in some places wood, enabled residents of Colorado to avoid the high cost of importing energy from other parts of the country.

Coal was Colorado's premier fuel. Readily available and accessible in many places, it heated homes and business establishments while serving as boiler fuel to generate steam for industry and transportation. It was the raw material for making coke, an essential fuel for the reduction of precious metal and iron ores. Coal, when distilled in retorts, yielded manufactured gas, which was distributed through underground pipes to urban homes and businesses.

The extraction and consumption of coal in Colorado began on a small scale in about 1860, and grew rapidly to become one of the state's major industries in the 1880s. Coal's preeminence among fuels influenced the Colorado General Assembly, in 1883, to create the Office of State Coal

Mine Inspector. Until then, there was no reliable information about the number and locations of mines, men employed, and annual production. With publication of the inspector's biennial reports, the first for 1884–1885, it is possible to follow the industry's growth with great accuracy.

Coal mining experienced alternating periods of prosperity and recession. In Colorado, as in the rest of the nation, there was a trend of rising coal consumption, which can be attributed to population growth and the expanding use of steam as a motive power in commerce and industry. Annual production grew at a steady pace until interrupted by a short but severe recession in 1884–1885. Expansion resumed until interrupted by a depression that began with the panic of 1893, the effects of which continued through 1897. Other recessions occurred in 1907–1908, and again in 1913 and 1914, a time of widespread labor unrest among coal miners. A recession in 1921–1922 marked the end of the coal industry's long-term growth.

Coal is, in fact, a series of carbonaceous fuels. They differ one from the other in their content of fixed carbon, volatile matter, and moisture. The higher the percentage of carbon, the greater the heat value. Anthracite ranks at the top, followed in descending order by bituminous and subbituminous coals, and lignite. All of these varieties were produced in Colorado.

Small deposits of anthracite were mined in the Uinta region, particularly in Gunnison County. The output, while never large, was readily marketed in urban centers throughout a large region because the fuel, when burned by residential and industrial customers, yielded little soot and smoke.

Bituminous coal, found in many places throughout the state, was produced in two grades. One, a metallurgical fuel that abounded in the San Juan, Raton, and Uinta Basins, was the raw material for making coke in beehive ovens, and later in by-product ovens. This eliminated most of the volatile matter and moisture, reducing the fuel to almost pure carbon, which was ideal for smelting precious metals and iron ores. Noncoking bituminous coal, which abounded in Huerfano County and in the northwestern quadrant of the state, was a superior boiler fuel. Railroads preferred it as locomotive fuel.

Subbituminous coal was mined throughout the Denver Basin, and at numerous places elsewhere in the state. It was usually found in shallow, thick seams that could be mined economically, and was widely consumed by households and industries. Subbituminous coal, because of its high moisture content, could not be transported long distances or stored for extended periods of time.

Lignite deposits were located on the eastern edge of the Denver Basin. Because lignite was low in carbon and high in volatile matter and moisture, it could not be produced and marketed in competition with the state's abundant superior grades of carbonaceous fuels. On at least two occasions, well-publicized plans to mine lignite in the vicinity of Scranton, in eastern Arapahoe County, were not carried to fruition.

Coal was produced in eight regions (see Map 1), which encompassed all or parts of thirty of Colorado's sixty-three counties. The Cañon City region, located in the south-central portion of Fremont County, produced a subbituminous coal that was an ideal boiler and domestic fuel. The original mines were located at Coal Creek, Rockvale, Brookside, and Williamsburg, between Florence and Cañon City. After the turn of the century, large-scale production shifted to what was called the south field.

The Denver Basin, stretching from Colorado Springs in the south to Fort Morgan in the north, produced subbituminous coal. Large amounts of this free-burning fuel were sold in the Denver market and to railroads. The two principal fields were Boulder-Weld and Colorado Springs. Northern mines were located at Marshall, Superior, Louisville, and Lafayette in Boulder County; and at Erie, Dacono, Frederick, and Firestone in Weld County. Mines near Colorado Springs supplied mainly a local market, including Cripple Creek and Colorado City.

The Green River region of Moffat and Routt Counties, an extension of the region of the same name in southwestern Wyoming, contained thick seams of bituminous and subbituminous coal. Commercial mining began in the Yampa field following the completion of the Denver and Salt Lake Railroad (Moffat Road) to Steamboat Springs in 1908, and to Craig in 1913. The North Park region, located in Grand and Jackson Counties, comprised two fields, North Park and Middle Park. Only the former produced coal; the latter contained deposits that did not warrant commercial development. The state's first commercial surface mines were in Jackson County.

The Raton Mesa region, the Colorado portion of which was located in Huerfano and Las Animas Counties, contained the Trinidad and Walsenburg fields. The former produced bituminous coal, and was the location of many beehive coking-oven plants. Colorado-made coke was widely used for the reduction of precious metals and iron ores in Colorado and throughout the region. The Walsenburg field produced noncoking coal, which was marketed as boiler and domestic fuel.

The San Juan Basin, extending northward from the Colorado–New Mexico border, encompassed portions of Archuleta, Dolores, La Plata, Montezuma, Montrose, Ouray, and San Miguel Counties. Coal was mined in four fields: Durango, Nucla-Naturita, Pagosa Springs, and Tongue Mesa. Historically, Durango produced coking as well as noncoking bituminous coals. Coke manufactured locally in beehive ovens was sold to regional smelters. The Nucla-Naturita, Pagosa Springs, and Tongue Mesa fields produced boiler and domestic fuel.

The South Park region, located in Park County, was active from 1870 to 1905. Some coal mined near Como was carbonized in beehive ovens for the smelting industry. The region's output, mostly subbituminous coal, was consumed as locomotive fuel by the Denver and South Park Railway, and for space heating by households and businesses in communities along that railroad.

One-half of the Uinta region occupied portions of seven Colorado counties. Eight fields have been active, mostly since the 1880s. These include the Book Cliffs (Mesa County), Grand Mesa (Delta County), Somerset (Delta and Gunnison Counties), Crested Butte (Gunnison County), Carbondale (Garfield and Pitkin Counties), Grand Hogback (Garfield and Rio Blanco Counties), Danforth Hills (Moffat County), and Lower White River (Moffat County). The Uinta region was an important source of anthracite as well as coking and noncoking bituminous and subbituminous coal. Coal was consumed on a small scale at the time of the Pikes Peak gold rush. Surface outcroppings were mined to provide fuel for people who lived along the Front Range of the mountains from Trinidad to Boulder. The quality of the fuel varied from one locality to another. It was consumed locally. The cost of wagon freight made it expensive at points distant from mines. High transportation costs and seasonal demand discouraged large-scale production.

Coal was mined commercially at an early date in Boulder County, northwest of Denver. Joseph M. Marshall, who came to the area in 1860 from Galena, Illinois, by way of Minnesota, discovered coal, iron ore, and limestone along South Boulder Creek. These were the ingredients for making pig iron. Marshall, with A. G. Langford, operated a foundry on Blake Street in Denver that melted scrap metal to make iron. Subsequently, they

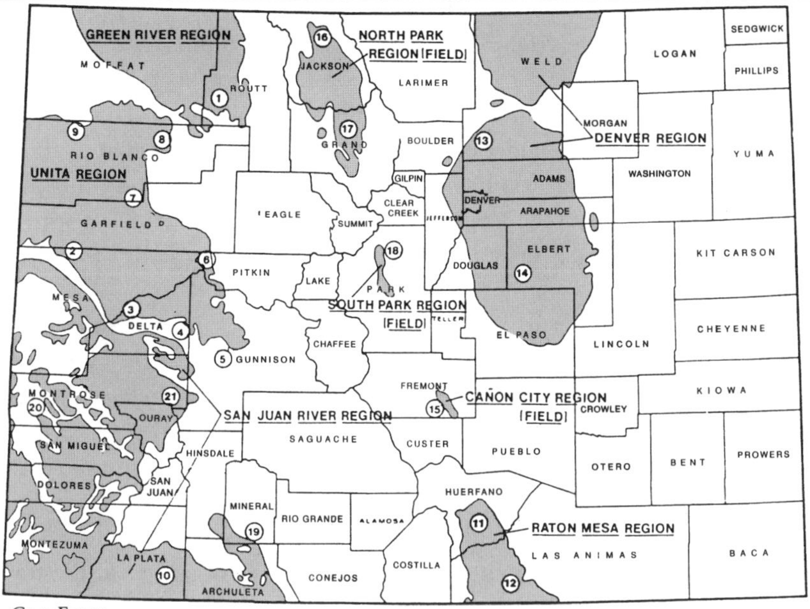

Coal Fields

1. Yampa
2. Book Cliffs
3. Grand Mesa
4. Somerset
5. Crested Butte
6. Carbondale
7. Grand Hogback
8. Danforth Hills
9. Lower White River
10. Durango
11. Walsenburg
12. Trinidad
13. Boulder-Weld
14. Colorado Springs
15. Cañon City
16. North Park
17. Middle Park
18. South Park
19. Pagosa Springs
20. Nucla-Naturita
21. Tongue Mesa

Map 1. Colorado Coal Regions and Fields. From Colorado Geological Survey, *Special Publication 41* (1996).

used raw materials found in Boulder County in an attempt to make pig iron. Lack of success caused Langford to move the foundry to the mining camp of Black Hawk in 1862, while Marshall developed coal lands in Boulder County. Under the name of Belle Monte Furnace Iron and Coal Company, he built a furnace and produced Colorado Territory's first pig iron in 1864. Fuel from the hamlet of Marshall was hauled to local ranches and to the emerging town of Boulder.

Obstacles to growth largely disappeared in the 1870s with the construction of iron rails to, from, and throughout Colorado. The Denver Pacific and the Kansas Pacific Railroads, Union Pacific (UP) subsidiaries that linked Denver with Wyoming and Kansas communities, respectively, sought locomotive fuel in Boulder and Weld Counties. Furthermore, by linking mines in those counties with the territorial capital, the UP hoped to tap a growing market for solid fossil fuel. The construction of the Denver and Boulder Valley Railroad, another UP subsidiary, from Hughes (later renamed Brighton) to Boulder encouraged the opening of numerous mines. When the initial segment was completed to Erie, in Weld County, in January 1871, the railroad company opened the Boulder Valley Mine. It was the largest producer of coal in the state's northern mining district for several years.

The Erie-to-Boulder segment of the line, completed in September 1872, had a similar impact on Canfield, immediately west of Erie in Boulder County. The community was named for Ira and Isaac Canfield, father and son, who had relocated from Pennsylvania to Colorado in the 1870s. They were originally from Livingston County, New York. The family moved to western Pennsylvania in 1861, where Ira and Isaac worked in the oil fields and coal mines, acquiring skills they would later apply in Colorado. In 1871, Ira, with his family, joined the Union Colony, organizer of the town of Greeley, the seat of government for Weld County. The Canfields bought several town lots and 105 acres of irrigated land on which they cultivated crops. Both father and son were attracted to land in Boulder County because of its potential for oil and coal. On finding the latter, in 1873 they opened the Rob Roy Mine adjacent to the tracks of the Boulder Valley Railroad, a source of subbituminous coal commonly known at the time as "black lignite." In 1879, following the sale of the Rob Roy Mine to the Jackson Coal Company, the Canfield family moved to Cañon City, Fremont County, where bituminous coal and petroleum had been produced for several years. There they played important roles in the further development of both resources.

The Union Pacific was the first of many railroads that were large purchasers of Colorado's coal for locomotive fuel. In the mid-1880s, railroads consumed 42 percent of the coal produced in the United States, and probably burned as fuel no less than that proportion of the coal mined in Colorado. In addition to consuming large amounts of coal, the railroads were vital links between mines and the state's growing urban industrial centers. In 1881, Denver required approximately 200,000 tons of coal a year, or about 600 tons daily, all of which reached the city by rail. In April of that year, the mines of the Cañon City district supplied 71,500 tons, followed by about 40,000 tons each from Weld and Boulder Counties. During the winter months, when the demand for fuel was high, the Marshall Mine near Boulder shipped an average of 3,174 tons of coal a month to the city.

The Union Pacific's mines along its rights-of-way in Colorado, as well as in Wyoming and Utah, filled the company's fuel requirements and allowed it to supply solid fossil fuel to communities throughout a large region. A land-grant railroad, the UP, by law, was entitled to develop the coal and "other minerals," undefined by terms of the grants, on the properties it received from the government in exchange for construction of tracks. By grant or by purchase, the railroad opened mines along its rights-of-way and, if necessary, built branch lines to gain access to economical and reliable sources of fuel.

Coal was a magnet that attracted other railroads to Colorado. For almost two decades they largely dominated the mining of solid fossil fuel. Large producers were the Atchison, Topeka, and Santa Fe; Denver, Texas, and Fort Worth; Chicago, Burlington, and Quincy; Chicago, Rock Island and Pacific; Colorado Midland; and Denver and Rio Grande. Because of this activity, Colorado became the leader among the six coal-producing states and territories of the intermountain West, and ranked among the top ten of the nation's twenty-eight producing states and territories in the 1880s and 1890s.

Inevitably, easy access to abundant minerals, including fuels, encouraged some railroads to organize heavy industries. The carriers needed rails and other iron and steel products, and freight revenues to be derived from the transportation of raw materials and finished products. An outstanding example of ancillary development was the Denver and Rio Grande Railway's iron and steel plant at Pueblo, for sixty years the only fully integrated metallurgical plant west of St. Louis.

The driving force behind the Denver and Rio Grande's industrial scheme was General William Jackson Palmer. A veteran of the Union Army, in which he had served with distinction, Palmer, at the close of the Civil War, was selected by the Kansas Pacific Railway Company to supervise the construction of the final segment of its line to Denver. When officials decided not to extend the line beyond that city in 1870, as originally planned, Palmer and a group of associates projected an independent line from Denver southward to Mexico. Supported by eastern and European capitalists, Palmer organized the Denver and Rio Grande Railway Company (D&RG) to build, in 1872, the first 100-mile segment of a narrow-gauge line to Colorado Springs and Pueblo.

In need of coal for fuel and for freight, Palmer and Dr. William Abraham Bell, an Englishman who had close ties to investors in his native land, formed subsidiary enterprises to assist in extending the narrow-gauge line from Pueblo westward up the Arkansas River Valley to coal banks near Cañon City, and southward to the Cucharas River, at the future site of Walsenburg, and on to El Moro, a new community four miles north of Trinidad. Walsenburg and El Moro were in the midst of rich coal deposits. Eventually, the D&RG extended its line to Trinidad, where Palmer's dream of reaching Mexico ended, because the Atchison, Topeka, and Santa Fe acquired exclusive control of the pass over Raton Mountain, which straddled the Colorado–New Mexico border.

Palmer and Bell pushed the narrow-gauge line westward from Pueblo to Cañon City, and beyond in the direction of Utah. Another line extended from Walsenburg over La Veta Pass to the San Luis Valley, where one thrust was toward New Mexico, to the south, while another continued westward. The Durango Trust, controlled by the railroad's executives, was instrumental in founding the new town of Durango, in the vicinity of which were large deposits of bituminous coal. Farther north, the Durango Trust underwrote the construction of a branch line from Gunnison to anthracite and bituminous coal deposits in the vicinity of Crested Butte. Coal and high-quality coke from that place provided the D&RG with locomotive fuel and freight revenues.

The Denver and Rio Grande Railway acquired valuable mineral resources in southern Colorado that Palmer and Bell hoped to utilize for making iron and steel. Mines near Trinidad produced bituminous coal that, when transformed into coke, was an ideal blast-furnace fuel for making pig iron. Iron ore was found on Grape Creek, in Fremont County, and limestone, needed for flux, abounded in the valley of the St. Charles River south of Pueblo. When the Grape Creek deposit proved to be contaminated with titanic acid, alternative supplies were found in the San Luis Valley and north of Salida, both on the narrow-gauge line and within reasonable distances of the proposed plant.

The Colorado Coal and Iron Company (CC&I), formed by the merging of the railway's ancillary enterprises, constructed an iron and steel works at Bessemer, on the Arkansas River, opposite the community of Pueblo. The plant, which began operation in 1881, consisted of a blast furnace, a Bessemer converter, rolling mills, and related facilities. The enterprise's principal products were rolled, cast, and forged products for the Denver and Rio Grande Railway and other regional carriers.

In addition, the Colorado Coal and Iron Company was for a time the region's largest producer and distributor of coal and coke. CC&I acquired large tracts of coal land in southern and western Colorado from the Denver and Rio Grande, from which it marketed domestic, steam, and metallurgical fuels, as well as coke. Eventually CC&I's leadership was challenged by a former Iowa capitalist, John C. Osgood, who had close ties to the Chicago, Burlington and Quincy Railroad. Osgood's Colorado Fuel Company acquired mines and erected beehive ovens in southern and western Colorado. Aggressive marketing, combined with mismanagement in the rival firm, enabled Osgood to dictate the terms for a merger of the two enterprises as the Colorado Fuel and Iron Company (CF&I) in 1892.

The new corporation owned both mines and ovens. The former included underground workings, hoisting machinery, tipples for sorting and loading coal into railroad cars, and an assortment of related structures. In addition, mining operations in remote locations required the commitment of funds to build and maintain towns to house workers. These company towns were administered by the superintendents of the local mines. They were often benevolent despots at best, and petty tyrants at worst. They dictated employment policies, and resisted efforts on the part of people inside and outside the towns to unionize the workforce.

The Colorado Fuel and Iron Company, as the principal supplier of fuel to the region's railroads and smelters of precious metal ores, became the undisputed leader of the West's expanding fuel trade. The corporation's standing in the market was further enhanced in the 1890s when the Colorado Midland, the Santa Fe, and the Chicago, Rock Island and Pacific Railroads sold or leased their Colorado coal properties to Osgood's firm in return for long-term fuel contracts at attractive prices. This allowed the carriers to devote their capital exclusively to transportation services and to avoid expensive and potentially troublesome peripheral activities such as coal mining.

Commercial mining required extraction of coal from underground deposits. Methods of gaining access to coal seams are depicted in Figure 1. They varied from place to place, depending on the character of the land and the depth of the coal seam. Vertical and slope shafts were widely used in Huerfano, Fremont, El Paso, Boulder, and Weld Counties. In the foothills and mountains, as in Las Animas County and west of the Continental Divide, drifts (also called adits or tunnels) were common. Surface, or open-pit, mines were not developed until the twentieth century.

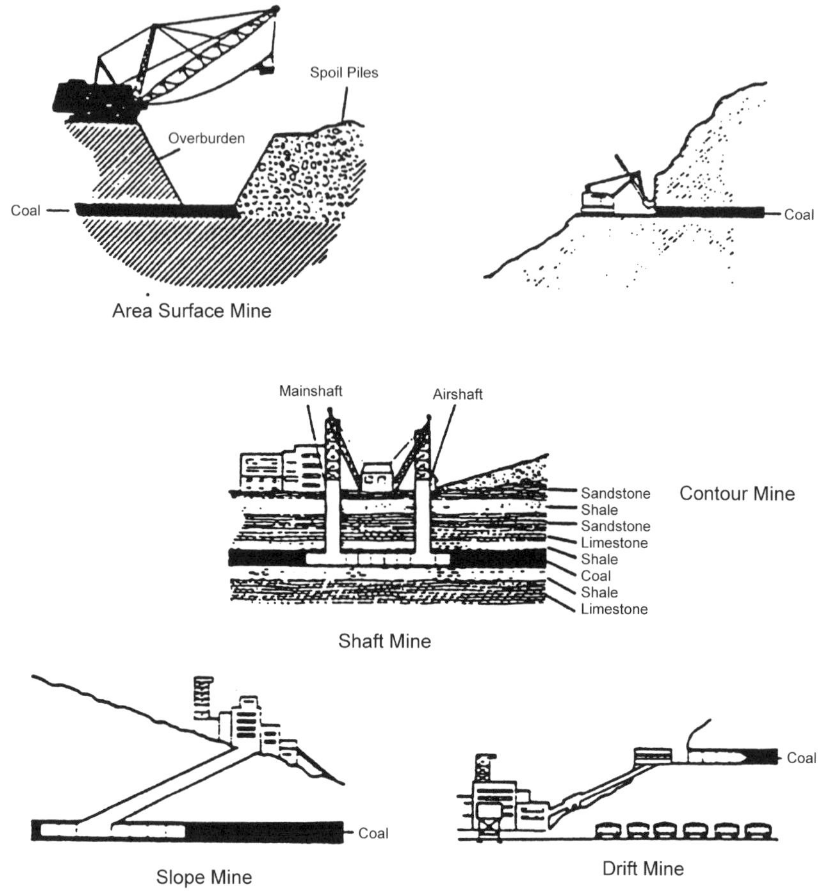

The method of mining a coal deposit depends on its depth and the character of the land.

Figure 1. Coal Mining Methods. From U.S. Energy Information Administration, *Coal Data: A Reference* (1995).

In the nineteenth century, coal mining was a labor-intensive activity. After shafts, slopes, and drifts reached the coal, main entries were driven into a seam, off of which rooms were opened. There a miner and a helper worked together to undercut the seam with picks, to drill holes with steel drills and hammers, and to pack them with explosive power that, on ignition, caused the coal to break away from the working face of the coal. It was then loaded into pit cars, each of about 1-ton capacity, and pushed along rails from the room to the entry, where men driving mules started them on their way to the surface.

The room-and-pillar method of extracting coal from seams was widely used in Colorado. Only one-half to two-thirds of the coal seam in a room was extracted. Pillars were left in place to support the roof and keep it from collapsing. The size of the pillars depended upon the condition of the roof. After all rooms had been mined, operating companies pulled or extracted the pillars as the final task before closing the mine. If roof conditions were bad, pillars were usually left in place, because the hazard of retrieving them was too great. In some fields, pillars had to be left in place to safeguard against surface subsidence.

Early-day coal miners were usually immigrants. In the 1870s a majority of mine employees were natives of the British Isles. As operations expanded, English, Scots, Welsh, and Irish, as well as American-born miners, were outnumbered by those who came from southern and eastern Europe. The new wave of immigration commenced with Italians in the 1880s, followed by people of Slavic origins—Poles, Slovaks, Slovenes, and Croats—then Greeks and other inhabitants of the Balkans. In the coal camps, people of Slavic origin were invariably known as Austrians, because at that time they came principally from the Austro-Hungarian Empire. Workers called Mexicans were, with few exceptions, U.S. citizens whose families had resided in the southwestern United States at the time of the Mexican Cession of 1848. Following a series of riots in the mid-1880s, Chinese were seldom found in Colorado's coal-mining camps, and other Orientals did not appear in large numbers until the first decade of the next century.

The influx of immigrants from southern and eastern Europe coincided with the rapid expansion of coal mining in the United States. In Colorado, the number of mines increased and the output of coal rose at a steady pace, along with the number of men employed in and about mines. By 1885, the office of the State Coal Mine Inspector reported fifty-three operating mines with a combined output of 1.4 million tons. Las Animas led all counties with 500,560 tons, most of which came from two properties, the Colorado Coal and Iron Company's Engle Mine and the Trinidad Coal and Coking Company's Starkville Mine. The latter was owned by the Atchison, Topeka, and Santa Fe Railroad. Fremont County ranked second with a combined output of 328,638 tons from six mines. Three properties—Coal Creek, Oak Creek, and Rockvale—accounted for 90.9 percent of the total. The first two were owned by the Colorado Coal and Iron Company, and the third by the Cañon City Coal Company, another subsidiary of the Atchison, Topeka, and Santa Fe Railroad.

Coal mining in 1885 was dominated by the two enterprises. The Colorado Coal and Iron Company's mines in Las Animas, Fremont, Huerfano, and Gunnison Counties had an aggregate output of 562,570 tons, or 40.2 percent of the state's total. The Santa Fe's mines in Las Animas and Fremont Counties reported a total output of 395,977 tons, or 28.3 percent of the total. Together, the two firms accounted for more than two-thirds of all the coal mined in the state.

In 1900, coal was produced in fifteen counties, as opposed to ten in 1885. Las Animas led with 2.2 million tons, or 40 percent of the state's total. It was, in fact, the only county to produce more than 1 million tons. Huerfano, Boulder, and Fremont, in that order, ranked second, third, and fourth.

Production in 1900 was controlled by a few enterprises. The CF&I, successor to the Colorado Coal and Iron Company, operated twenty mines with a combined output of 2.9 million tons. These included former Santa Fe mines operated under lease. CF&I's principal rival was the Victor Fuel Company with six mines and an aggregate output of 749,761 tons. The Northern Coal and Coke Company's fourteen mines had a combined output of 705,063 tons. The Union Coal and Coke Company's four mines reported a total production of 182,620 tons. Together, these four enterprises operated forty-four mines and accounted for 82.9 percent of the state's production. CF&I alone mined 53 percent of the state's output in 1900, far more than any of its rivals.

The CF&I grew by acquiring existing mines and by opening new ones. It purchased the Colorado portion of the Maxwell Land Grant in 1893, adding to its holdings thousands of acres of rich coal lands. Three years later, CF&I leased the Colorado coal properties of the Atchison, Topeka, and Santa Fe Railroad, a total of 20,000 acres that included, in addition to the Starkville and Rockvale Mines, the Vulcan Mine at New Castle, in Garfield County.

The Victor Fuel Company's five mines in Las Animas County and one in Fremont County contributed 14 percent of Colorado's production in 1900. Organized in 1899 as the successor to the Victor Coal and Coke Company, the corporation was controlled by John C. Osgood, who, as president of the CF&I, hoped to merge the fuel operations of the two firms. However, before he had the opportunity to do so, he suffered financial losses in 1903 that forced him to give up control of CF&I to George Jay Gould and John D. Rockefeller. Osgood retained the Victor firm, and later merged it with the American Fuel Company, operator of mines at Gallup, New Mexico. The combined enterprise was known as the Victor-American Fuel Company.

The Northern Coal and Coke Company, with mines north of Denver, claimed 13 percent of Colorado's output. The corporation had been organized by James Cannon Jr. with financial support from eastern investors. He had been president of the United Coal Company, which at one time owned eleven coal properties, six of which were in Boulder County. When that firm was forced into bankruptcy in 1896, Cannon quietly bought up its outstanding bonds and used them to purchase from the receiver all of United's properties. He created the Northern Coal and Coke Company by combining United's assets with those of the smaller Northern Coal Company, and set out to build an enterprise capable of commanding a large and profitable market share in an industry plagued by chronic excessive capacity. Too many producers, especially in the northern field, caused the supply of coal to outstrip the largely seasonal demand for it, with the result that most companies did not earn profits from the production of coal.

Excessive capacity undermined prosperity in most of Colorado's coal fields. The boom of the 1870s slowed from time to time, but continued throughout the first two decades of the twentieth century. The state's output of all grades of coal peaked at 12.1 million tons in 1910, then declined gradually to 8.7 million tons in 1915. At the same time, the average number of men employed in the mines fell from 14,768 to 12,563. The contraction ended in 1916 because of an abnormal demand for American goods triggered by World War I in Europe. In 1917–1918, as a participant in that conflict, the United States mobilized its manpower and industry to aid the Allies in defeating Germany and the other Central Powers. Wartime activities boosted coal output to a record 12.7 million tons in 1918, and to a near-record of 12.5 million tons in 1920.

The coal-mining recession of 1910–1915 created widespread dislocation in the industry. Mine owners and operators responded to falling demand by adopting machines to supplant costly hand methods of extracting coal. Lower costs of production could, in theory, reduce selling prices and enlarge a company's share of the fuel market. At the same time, mine workers sought to organize a statewide union as an instrument for avoiding, or at least softening, the impact of layoffs and changes in methods of producing coal. This set the stage for a power struggle in 1913–1914 between large mining companies, which tolerated no interference with their right to respond, as they saw fit, to changing market conditions, and the United Mine Workers of America (UMWA), which wanted to be the collective-bargaining agent for all Colorado mine workers.

Labor unrest was not unusual in Colorado's coal mines. In Boulder, Weld, Fremont, and Las Animas Counties, employees had organized in the 1870s and 1880s to gain better wages and working conditions. The first attempt to gain recognition for statewide collective bargaining, led by the Knights of Labor in 1884, was successfully resisted by the mining companies. Continuing resentment over a system that denied miners a voice in determining conditions under which they labored in mines prompted frequent strikes, which were almost annual occurrences in some areas. Walkouts were seldom successful, for miners, as individuals or groups, could not match the economic and political power of the mine owners.

Another general strike in June 1894 closed most of the state's mines. The workers wanted increased wages, regular paydays, and an end to wages paid in scrip, which was redeemable only in company stores. Scrip had become an irritant because cash-starved firms, struggling to continue operations in the depression that followed the panic of 1893, had no other way to compensate workers. There were numerous demonstrations, but almost no acts of violence. The operators refused concessions, and by early fall the men returned to work.

The strike of 1894, followed two years later by a troublesome walkout of Leadville's hard-rock miners, inspired the Colorado General Assembly to authorize a state board of arbitration to avoid strikes and lockouts by peaceably resolving differences between employers and employees. The law, adopted in 1897, empowered the board, on receipt of written requests from either side in the dispute, to recommend terms for settling the dispute. However, the board's findings could be ignored by either party.

The Board of Arbitration enjoyed some small successes, but failed to resolve major disputes. A case in point was the strike by 350–400 miners that closed the Northern Coal and Coke Company's mines at Erie, Lafayette, and Louisville commencing in December 1900. In the course of the walkout, the Board of Arbitration, on request from miners, examined all issues and made recommendations for a settlement that incorporated most of the strikers' demands. The Northern company, however, refused to accept the recommendations and the strike continued. In May 1901, fearing that the company was preparing to reopen the mines with imported labor, the strikers accepted terms offered by the employer, including a 10 percent increase in wages and a reduction in the cost of powder at the company stores. The company refused to recognize the UMWA as a bargaining agent for workers. The five-month disruption of operations undermined the Northern company's competitive position in the coal trade, and stymied its long-range plan for stabilizing production in Boulder and Weld Counties.

Two years later, the Northern Coal and Coke Company sidestepped a prolonged strike. The miners asked for higher wages, an eight-hour workday, biweekly paydays, compensation for all coal mined, and an end to discrimination against members of the UMWA. The strike commenced October 26, 1903; within three weeks, F. F. Struby, an official of the Northern company, speaking for all operators in the field, offered an eight-hour day and increases in wages. The miners accepted the offer on November 28, and 1,270 men returned to work. The mines north of Denver were active for the next several months while a strike dragged on in the southern field. The Colorado Fuel and Iron Company and the Victor Fuel Company called on local sheriffs and the state militia to protect workers who were imported to keep the mines active, with the result that the strike ended, for all practical purposes, by July 1904.

Colorado's coal industry remained relatively free of labor problems for a few years. Operators in the northern field recognized the United Mine Workers of America in 1908 as collective-bargaining agent for employees. Labor peace lasted until 1910, when employers unilaterally abrogated their agreement with the union, insisting on wage reductions and other changes in long-standing policies. On March 1, about 3,000 miners in Boulder and Weld Counties walked off their jobs, calling for a union contract that included wage increases and other benefits.

Under the leadership of F. F. Struby, by then president of the Northern Coal and Coke Company, Boulder and Weld County mine operators refused to submit the miners' demands to arbitration, but quickly arrived at a compromise that reopened mines. Four companies—Northern Coal and Coke, American Fuel, Louisville Coal and Land, and the Big Six—resumed production, while five others—Rocky Mountain Fuel, National Fuel, Frederick Fuel, Brooks-Harris Fuel, and Consolidated Coal and Coke—

moved forward with plans to replace the strikers with nonunion men. The walkout dragged on for more than four and a half years, the longest coal miners' strike in Colorado history.

Ironically, the Northern Coal and Coke Company was an early victim of the labor unrest it sought to avoid. Declining cash flows forced the enterprise to dispose of its properties for approximately $1.5 million. On October 16, 1911, the Rocky Mountain Fuel Company bought from its former competitor ten mines, seven company stores, and coal land, mostly in Boulder County. The new owner took possession one week later.

The Rocky Mountain Fuel Company's acquisitions, because of its operations in the south, transformed the labor unrest in mines north of Denver into a statewide strike. In September 1913, the United Mine Workers of America, in the hope of strengthening the walkout in the northern field, called out miners in Fremont, Huerfano, and Las Animas Counties. The strike spread also to the Western Slope, particularly to Gunnison, Routt, and La Plata Counties.

The UMWA hoped to win concessions from the state's most powerful coal-mining enterprises—the Colorado Fuel and Iron Company, the Victor-American Fuel Company, and the Rocky Mountain Fuel Company. Demands included a 10 percent advance in wages, the right of workers to trade at mercantile outlets other than company-owned stores, strict enforcement of state mining laws, abolition of mining-camp guards, and recognition of the union as the collective bargaining agent for all coal miners in Colorado.

About 7,600 men, probably three-fourths of the workforce in Fremont, Huerfano, and Las Animas Counties, responded to the union's call. The strikers vacated their homes in company towns. Trinidad district miners and their families found refuge in tent communities, or colonies, erected by the UMWA at Ludlow, Primero, Segundo, and Sopris. The largest was at Ludlow, a strategic location at the head of the canyons leading to the area's principal mines. This afforded strikers an important advantage for taking actions to discourage strikebreakers imported by the companies from taking jobs in the mines.

The southern counties became an armed camp. Each faction assumed that the other was prepared to use force to achieve its objectives. Numerous, often violent, skirmishes, usually in the vicinity of the tent colonies, persuaded Governor Elias M. Ammons that local authorities could not maintain law and order. He ordered National Guard units to the strike zones in October 1913. Peace prevailed as long as the military was on the scene, but the government, lacking funds to support the units, withdrew most of them in April 1914. Taking their place were troops recruited specifically for strike duty, mostly men who were sympathetic to views of the coal companies.

On April 20, 1914, the area around Ludlow erupted in gunfire. A pitched battle between the state troops and the strikers continued throughout most of the day. One militiaman and five miners were killed, and others were wounded. During the battle the tents of the Ludlow colony burned to the ground. Later it was discovered that two women and eleven children, who had taken shelter in shallow pits under the floors of the tents, had suffocated during the fire.

Each side blamed the other for the tragedy at Ludlow. In fact, it was not clear who started the skirmish at Ludlow, or who set fire to the tents. The UMWA immediately labeled the events of April 20 a "massacre," attributing blame to the coal companies and the militia. The tragedy focused public attention on the strike, and on labor relations in the southern field.

The events at Ludlow inflamed striking miners throughout Colorado, prompting attacks on company properties. A reign of terror continued for several days. When additional state troops could not return peace to the coal fields, President Woodrow Wilson dispatched units of the United States Army to end the disturbances. Federal troops disarmed both sides, but that did not bring an end to the dispute.

The strike dragged on for another six months. The mine operators refused to negotiate a settlement. The UMWA, its funds exhausted, finally declared the strike at an end in December 1914. Most of the strikers returned to work without having gained any concessions from their employers. The bitterness generated by the long strike precluded the quick restoration of much-needed harmony in labor-management relations.

In the aftermath of the strike, John D. Rockefeller Jr., whom the public blamed for the events at Ludlow, acting as the majority stockholder in the Colorado Fuel and Iron Company, introduced an Employee Representation Plan, widely known as the Rockefeller Plan, in an effort to restore harmony to employee-employer relations. It created a company union designed to give workers a voice, although a minor one, in the determination of wages and conditions of labor. In addition, a program of welfare capitalism provided workers with improved housing, health care, recreation facilities, and educational opportunities. A majority of the workers voted for the plan, and it remained in place for almost twenty years, but it did not achieve harmony in the mining camps, and it did not end periodic strikes. However, there was no repetition of the extreme violence of 1913–1914.

Steelworkers, but not coal miners, enthusiastically supported the Rockefeller Plan. It failed to gain the long-term support of miners largely because stagnation gripped the coal industry. Production in Colorado, after peaking in 1918 and 1920, declined during a long-term recession. The larger companies were hard hit by the recession. Near-record production in 1920 convinced many operators that they had escaped the much feared postwar depression. They were wrong. In 1921 and 1922 the United States experienced a short but severe economic adjustment, during which coal production dropped dramatically in Colorado, as it did throughout the nation. The recession ended in less than a year, after which coal mining, unlike some sectors of the economy, did not experience full recovery. The industry suffered from a familiar problem—excessive capacity—in that it produced far more coal than could be marketed at prices that gave to mine owners fair returns on their investments and efforts. Stated simply, Colorado, along with other states, produced too much coal. Alternative fuels were available at attractive prices. Consumers turned from coal to petroleum and natural gas in large numbers. The remedy was to bring the output of coal in line with demand, but mine owners, if they recognized both the cause and the cure for the problem, were reluctant to initiate appropriate remedies.

Modest increases in output occurred in 1922, and continued for another four years. Thereafter, production declined sharply. This was largely due to increased competition from other fuels. Excessive capacity plagued the industry for the remainder of 1920s, and was exacerbated by the Great Depression, which commenced in 1929 and continued for more than a decade.

The decline of coal consumption was a long-term phenomenon. The solid fossil fuel had supplied more than 90 percent of the nation's energy needs in 1890, but its share of the market had fallen to 77 percent in 1910, and to 73 percent in 1920. This downward trend continued, even accelerated, in the 1920s. At the end of the decade, coal retained only 58 percent of the fuel market. Petroleum and natural gas had claimed only 5 percent of the market in 1900, but made fast gains at the expense of coal in the 1920s, increasing their combined share to 32 percent in 1930.

Prior to 1910, Colorado's principal mines had been located along the Front Range of the Rocky Mountains from Boulder, in the north, to Fremont, Huerfano, and Las Animas Counties in the south. Gunnison County was an important source of fuel for the smelting and gas-making industries. By 1920, mines in Routt County, in the northwest, and Weld County, in the Denver Basin, captured a large share of the fuel market. Routt County's superior quality of bituminous coal gained access to Denver and other in-state and out-of-state markets with the completion of the final segment of the Denver and Salt Lake Railroad, known as the Moffat Road, from Steamboat Springs to Craig in 1913. Weld County mines produced a superior grade of subbituminous home-heating and boiler fuel, which could be mined from thick seams cheaply and in quantity by machine methods. They enjoyed a competitive edge over older mines in Boulder County.

The impact of the recession of the 1920s may be seen in a comparison of conditions in 1920 and 1930. In the former year, coal mining was riding the crest of prosperity resulting from the country's participation in World War I. That year, 231 mines in Colorado employed 13,665 men an average of 256 days to produce 12.5 million tons of fuel. The fuel was widely consumed in heating homes and businesses, and as boiler fuel to generate steam for transportation, manufacturing, and electric utilities. By 1930, however, the state's 275 mines, an increase of more than 10 percent since 1920, employed only 10,015 men an average of 165 days to produce 8.2 million tons of coal. In the 1920s, as mines had closed or reduced their workforces, many unemployed miners opened small wagon or truck mines in the hope of supplying local markets.

The "big three" companies absorbed the major impact of the recession. In the 1920s the Colorado Fuel and Iron Company's operating properties shrank from twenty-two to eleven, the Rocky Mountain Fuel Company's from sixteen to six, and the Victor-American Fuel Company's from nine to four. At the same time, CF&I's output fell from 4.2 million to 1.8 million tons, while the Rocky Mountain Fuel Company's production declined from 1.2 million to 0.7 million tons, and Victor-American's from 1.1 million to 0.7 million tons.

A severe blow to the Colorado industry was the substitution of fuel oil for coal by the region's railroads, and the importation of natural gas, commencing in 1928, from Texas to supply homes, businesses, and industries along the Eastern Slope. The Colorado Fuel and Iron Company, the state's largest heavy industry as well as its major fuel enterprise, adopted natural gas in its steelworks because it provided significant savings over coal. In making this change, CF&I eliminated the need for up to 2.5 million tons of coal a year from its own mines, mostly in southern Colorado. Similarly, the adoption of natural gas for heating homes and factories in Pueblo, Colorado Springs, Denver, Fort Collins, and smaller communities eliminated a large market for domestic coal. This loss had a dramatic impact on Boulder and Weld Counties in the northern field and Fremont County in the south, the principal sources of coal for the home-heating market.

Competition from oil and natural gas forced mine operators to reduce the price of coal by lowering the cost of production. To this end, they further abandoned traditional methods for extracting coal. Machines rapidly displaced hand-labor methods of mining. Even before the turn of the century, steam-powered air compressors had been introduced into some mines, permitting the use of drills and undercutters. Improved models were widely used by the 1920s, while, to a large extent, compressed-air devices were superseded by electric-powered equipment. Electricity, an economical and flexible source of energy, hastened the application of machinery to all aspects of mining.

Notable advances were mechanical methods of loading and more efficient ways of moving coal from underground workings. Pit-car loaders were adopted in the 1920s, displacing men with shovels at the working face of the coal seam. Self-propelled, tire-mounted loaders and belt-conveyers to move coal from underground workings to the tipple were not widely adopted until after World War II, but loading became the measure of mechanization in underground mines.

Mechanization exacerbated the problem of excessive capacity. Because the ability to produce already far exceeded demand, coal mining was generally recognized as a "sick industry" in the 1920s. In Colorado, mines closed and workforces were reduced, but not fast enough to adjust to the shrinking market. The social consequences of unemployment and the abandonment of mining towns were far-reaching, but less unsettling than the poverty that resulted from underemployment for the men, and their dependents, who eked out a meager living at best working two or three days a week rather than the previously normal five or six. In addition, those who remained in the mines faced the threat of closings or displacement by machines.

Disgruntled employees had traditionally turned to the UMWA for redress of grievances, but that once-powerful union, in disarray because of

The Columbine, at Serene, near Erie, in Weld County, was the Rocky Mountain Fuel Company's largest mine in the northern field. During the IWW strike of 1927–1928, the Columbine Mine was the scene of a clash between miners and the Colorado Rangers, resulting in deaths and injuries. In the aftermath of that tragedy, the Rocky Mountain Fuel Company, in an attempt to achieve harmony between operators and miners, broke rank with the owners by recognizing the United Mine Workers of America as the bargaining agent for employees. Courtesy, Erie Historical Society, Erie, Colorado.

the lingering effects of the strike of 1913–1914, was powerless to deal with problems arising from the shrinking market for coal. In desperation, miners in Colorado and adjacent states turned for help to the Industrial Workers of the World (IWW). A union with a long history of radicalism, the IWW called for an end to capitalism, and substitution for it of an industrial system controlled by workers. This utopian system was the goal of the IWW cadre, who led Colorado strikers in an attempt to overthrow the existing system in 1927–1928.

The walkout affected all producing fields, as pickets sought to halt coal production throughout Colorado. Predictably, the use of force led to counterforce, with the police power of local and state governments aligned on the side of the mine owners and operators. Violence was widespread, but the loss of lives was avoided until heavily armed Colorado Rangers, successors to the militia for quelling strikes, clashed with strikers at the Columbine Mine near Erie in Weld County, on November 21, 1927. Six miners were killed, and more than sixty injured.

The Columbine Mine was the Rocky Mountain Fuel Company's largest property. The head of the enterprises had been John J. Roche, vice-president from 1914 to 1922, and president until his death in early 1927. Roche's only child, Josephine, had earned a master's degree in social work at Columbia University in 1912, after which she served for a time as Denver's first female police officer. As the principal stockholder in the Rocky Mountain Fuel Company, Miss Roche, shocked by the violence of the strike, set as her goal improved relations between the firm and its employees.

Determined to change the Rocky Mountain Fuel Company's labor policies, she secured, in 1927, the election of Merle D. Vincent, a Denver lawyer, as vice-president and general manager. Vincent immediately broke ranks with other mining enterprises by recognizing the United Mine Workers of America as the collective-bargaining agent for all of the firm's employees. The contract with the UMWA was the only one at that time in Colorado.

There was no relief from conditions that caused the strike of 1927–1928. The recession that plagued the coal industry in the 1920s deepened with the start of what came to be known as the Great Depression in 1929. Conditions for miners and their dependents deteriorated as economic stagnation intensified, and for a time there was general fear of the coal industry's total collapse. That did not happen, but it was not until well into the 1930s that improvement, however meager, finally occurred.

2

Smelter Fuel 1877-1930

In the 1870s the manufacture of coke became a thriving industry in southern Colorado, where coal suitable for making smelter fuel was mined on a large scale. Oven operations expanded in Las Animas and spread to Gunnison, Garfield, La Plata, and Pitkin Counties. In 1918, coke-making shifted largely from coal fields to the principal industrial site where the fuel was consumed, but the sources of metallurgical coal remained unchanged.

The first reduction furnaces to extract gold and silver from complex and compound ores utilized charcoal as fuel. At Leadville, one of Colorado's early smelting centers, furnaces used large quantities of the fuel made by carbonizing wood. The Grant smelter consumed 125,000 bushels monthly in 1881. Three plants used an average of about 100,000 bushels each month, and others burned smaller amounts. At that time, the price of charcoal varied from 12 to 14 cents a bushel.

Charcoal was readily available and relatively inexpensive. In the mountain district, at least for a time, abundant timber could be carbonized economically by burning it with limited access to oxygen in conical or domed kilns. However, during the ore reduction process, charcoal often produced uneven heat, creating what were called "hot tops." When that occurred, there was a large loss of molten metal. In addition, furnace charges were small because charcoal lacked the structural strength to support a heavy burden of ore and fluxing material. Smelter operators substituted coke, made from coal, for charcoal, if they could, because the greater efficiency of furnace operations more than offset the higher cost of the fuel.

Some western furnace operators turned to coke in the early 1870s. At that time it was available only from western Pennsylvania, and transportation costs were very high. Named for the district where it was produced, Connersville coke was almost pure carbon, and it produced little ash. Furnace yields from high-grade ores at the time justified the cost of coke, but in the hope of increasing profits, operators looked for sources in the West. They found what they wanted in the Raton Basin of southern Colorado and northern New Mexico. Mines near Trinidad, in Las Animas County, produced metallurgical coal from which a satisfactory grade of coke could be made cheaply in large quantities. Quality coke was also made on a small scale in New Mexico and in eastern Utah.

A hard, compact, porous gray mass of carbon, Trinidad-area coke proved to be an ideal fuel for reducing precious-metal ores. It was also used to reduce iron ore to pig iron, the basic material for making steel. When charged in furnaces with ore and flux, coke produced the high temperatures required to melt metals and to fuse impurities into slag. In addition, coke, unlike charcoal, had a dense structure capable of producing even heat throughout the furnace, while supporting heavy charges of ore and flux during the reduction process.

The potential demand for coal and coke by the West's expanding precious-metals mining and reduction industries was one of the reasons why General William Jackson Palmer and his associates organized the Denver and Rio Grande Railway to build southward from Denver to El Paso, Texas. After reaching Pueblo via Colorado Springs, the narrow-gauge line constructed a branch up the Arkansas River Valley to coal banks near the present site of Florence in Fremont County. The mines yielded a superior grade of fuel ideal for industrial and domestic heating, but not for making coke.

In pursuit of coking coal, the railway's promoters extended the narrow-gauge line from Pueblo southward to the Cucharas River, and on to the new town of El Moro, four miles north of Trinidad. On the Cucharas they found excellent coal for firing boilers, but it did not make satisfactory coke. Farther south they found large deposits of coking-grade bituminous coal. The D&RG organized a subsidiary, the Southern Colorado Coal and

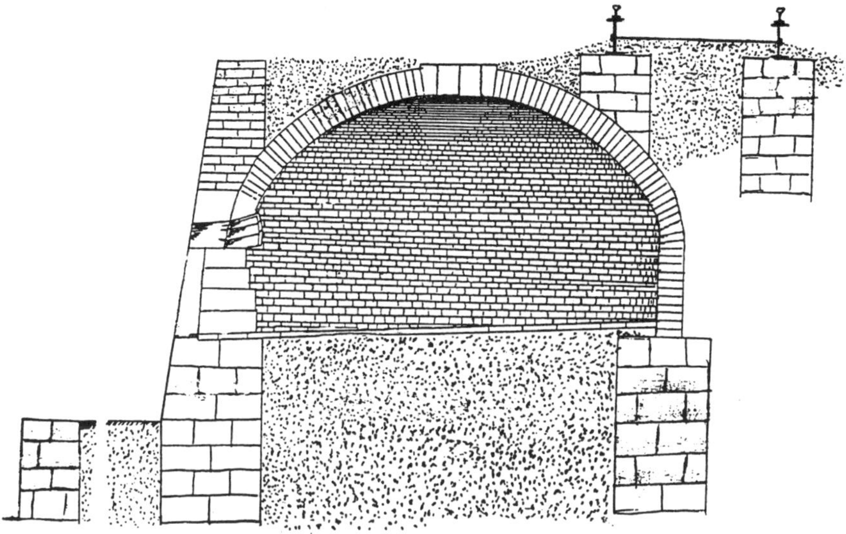

Figure 2. Cross Section of Beehive Oven. Adapted from *Camp and Plant* 5 (January 23, 1904): 31.

Town Company, to open a mine and to carbonize coal in newly constructed ovens at El Moro. George U. Engle, an engineer with previous experience in the mining and coking industries of Pennsylvania, supervised the construction of beehive ovens in 1877, most of the work being done by Italian immigrants and local residents.

Figure 2 is a cross section of the beehive oven used by Colorado's principal manufacturers of coke. The name was derived from its shape, which resembled the domed hives used by beekeepers in ancient times. Ovens were usually built side by side and back to back, forming batteries, with each unit measuring about 13 feet in diameter. The space between the ovens was occupied by stone retaining walls and piers, on top of which wooden (later steel) rails supported "larry" cars used to charge the ovens.

The tunnel head at the top of the dome allowed the oven to be filled with coal from cars moving along overhead tracks. The door at the left was closed, except for a small hole to admit air during the coking process. The carbonized coal, or coke, was pulled from the oven onto the wharf, where it was cooled and, if necessary, stored until shipment.

Coke-making was a lengthy process. Crushed coal from the El Moro Mine (later renamed Engle), six miles south of the ovens, was loaded into larry cars that one man, using the force of gravity, moved along the rails on top of each battery of ovens. After an appropriate amount of coal had been dropped through the funnel, or tunnel head, of each oven, men with scrapers leveled the charge and sealed the door with bricks, leaving one small air hole at the top. The heat of the oven from the previous charge baked the coal, causing the discharge of gases that ignited above the coal, allowing the heat of combustion to radiate downward for a period of forty-eight to seventy-two hours. Smoke from the ovens carried tar, ammonia, benzol, and other products liberated by combustion. These pollutants killed all vegetation within miles of the plant. The length of the process determined the structure of the finished product.

After carbonizing was complete, the bricks were removed from the door, through which workers sprayed the coke with water to extinguish any flames. Quenching had to be done with great care to avoid unnecessarily cooling the oven, in order to preserve heat for the next charge. After the coke had "steamed off" for a few minutes, men using ravels or scrapers pulled the charge onto the platform where, if necessary, additional quenching occurred. Tined implements called "forks" were used to load the coke into railroad cars for shipment.

Oven-tending, one of the least desirable jobs in the coal fields, was hot, dirty work. Recent immigrants were assigned to the ovens, and they were looked down upon by other ethnic groups. In the late nineteenth century, most jobs in and about the ovens were held by southern Italians and "Austrians," men of various Slavic origins.

The first shipment from El Moro, in February 1878, was to the Harrison Reduction Works at Leadville. The following year, fifty new ovens raised the total to 222 beehive and 28 stack ovens, having a combined output of 300–350 tons of coke per day. The stack ovens, a variation on what was called the Belgium design, were rectangular in shape, and proved to be less efficient than the beehive model. A washer, added in 1879, removed dirt and debris from coal before it entered the ovens. This reduced the ash content of the coke to levels requested by smelter operators.

El Moro coke was uniformly hard, cellular in structure, and, according to company officials, comparable, except in price, to Connellsville coke, the recognized benchmark for quality. Initially the El Moro product sold for $9.50 a ton, while the eastern variety cost $12.50 a ton delivered in Pueblo. An excellent product at a competitive price ensured the dominance, at least for a time, of El Moro coke in the western market. As the owner of the region's first center for the manufacture of smelter fuel, the Southern Colorado Coal and Town Company appeared to have a bright future. In Colorado alone there were, in 1878, more than five dozen silver-lead (argentiferrous) reduction works, most of which wanted an economical alternative for either charcoal or Connellsville coke.

Another market for coke appeared in 1881, when the Colorado Coal and Iron Company constructed an iron and steel works at south Pueblo. The plant's blast furnace consumed El Moro coke to make pig iron. The Colorado Coal and Iron Company (CC&I) was formed in 1879 by a consolidation of the Southern Colorado Coal and Town Company, the Colorado Improvement Company, and the Colorado Steel Works Company, all affiliated enterprises of the Denver and Rio Grande Railway Company. CC&I operated the region's only integrated iron and steel plant, and led all rivals in the production and distribution of coal and coke.

El Moro was the only source of Colorado-made coke until the 1880s. With the construction of additional ovens, El Moro's output rose steadily, exceeding 100,000 tons for the first time in 1882. That year the product claimed 85.1 percent of the state's output. Competition for the ore-reduction fuel market explains, at least in part, El Moro's relative decline during the balance of the 1880s and into the 1890s. The sporadic operation of the steelworks at Pueblo was another factor. The recession of 1884–1885, and the panic of 1893, with an ensuing depression, undermined smelter operations and adversely affected the production of pig iron. By 1900, El Moro accounted for 22.8 percent of the coke produced in Colorado.

Competitors were attracted by the prospect of large profits. In 1875 the Atchison, Topeka, and Santa Fe Railway Company, in anticipation of completing its transcontinental line from La Junta to Trinidad, created a subsidiary enterprise to acquire coal lands south of the latter community. The Trinidad Coal and Coking Company opened at the Starkville mines, which provided the carrier with fuel for its locomotives and bulk freight for distribution to points along its line in Colorado, Kansas, and New Mexico. In 1881, the Trinidad Coal and Coking Company erected beehive ovens at Starkville. That year, coke shipments began on a small scale to smelters, mostly in Arizona. By then, the price of coke had declined to about $5 a ton, where it remained for a number of years.

The Trinidad Coal and Coking Company's output fluctuated from 20,000 to 30,000 tons of coke a year in the 1880s, and from 40,000 to 50,000 tons a year in the early 1890s. The firm continued in operation until August 1896, when the Santa Fe Railway Company leased the mine and 140 ovens to the Colorado Fuel and Iron Company (CF&I), successor to the Colorado Coal and Iron Company. The lessee erected an additional 50 ovens and a washery to reduce the ash content of the coke. In 1900, the 190 ovens produced 90,493 tons of coke, which was sold to smelters in the region.

At Sopris, west of Trinidad and only a few miles from Starkville, John C. Osgood's Denver Fuel Company opened mines in 1887. One hundred beehive ovens began operation the following year, producing coke with a dense cellular structure so low in ash, sulfur, and phosphorus that it became the regional standard for smelter fuel. The Denver Fuel Company produced 15,096 tons of coke from September 1, 1888, to March 31, 1889. Its successor, the Colorado Fuel Company, also controlled by Osgood, turned out 147,068 tons of coke from April 1, 1889, to October 31, 1892, when it became part of the CF&I. That firm built an additional 122 ovens in 1893, and another 50 ovens in 1900, giving the Sopris plant a total of 272. In 1900, the plant produced 106,772 tons of coke. Some of it was consumed in the blast furnaces at Pueblo, but most was sold to smelters of precious metal ores in Colorado and elsewhere in the West.

The three principal centers for ore reduction in the Centennial State were Leadville, Pueblo, and Denver. The Pueblo and Denver plants consumed large amounts of coke produced in Las Animas County, but the high cost of shipping it from Trinidad to Leadville, a deterrent to the growth of mountain-based smelting activity, persuaded officers of the Colorado Coal and Iron Company to experiment with coke-making at Crested Butte. It was made in open pits as early as November 1881, but that method was abandoned when fifty beehive ovens were completed in March 1883. The product of those ovens had a well-developed cellular structure and less than 5 percent ash, which made it, in spite of relatively high transportation costs, an ideal fuel for smelters at Leadville. Another fifty ovens were erected in 1885, and an additional fifty-four by 1900. From 1881 to 1900, the Crested Butte ovens turned out 805,833 tons of coke, or an average of 40,292 tons a year.

The Sopris beehive ovens, southwest of Trinidad, Las Animas County, Colorado, ca. 1890, used man- and mule power. The mule-driven larry car dropped the proper charge of coal into each oven. The twelve-man crew leveled the charge and closed the oven door, except for a small air hole at the top. Twenty-four hours later, they quenched the flames and pulled the coke onto the platform. After it had cooled, they loaded it into railcars for transportation to smelters and the iron furnaces at Pueblo. Coke-making ended at Sopris in 1918. Courtesy, Bessemer Historical Society, Pueblo.

In addition to Crested Butte, the Leadville smelters acquired coke from the Grand River Coal and Coke Company, operator of ovens at Cardiff, located on a branch of Colorado Midland Railway four miles south of Glenwood Springs, in Garfield County. That enterprise had its origin in the efforts of Jerome B. Wheeler, principal owner of the Aspen Mining and Smelting Company, to find a local source of coke for the Aspen Smelter. After unsuccessful experiments with coal mined in Jerome Park, south of Glenwood Springs, Wheeler entered into an agreement with James J. Hagerman, president of the Colorado and Midland Railway, to surrender extensive coal lands in Garfield and Pitkin Counties to a new enterprise, the Grand River Coal and Coke Company, which was to be capitalized at $1 million to mine coal and manufacture coke for the Aspen, Leadville, and Denver markets. That corporation, chartered in New York on May 12, 1886, opened four mines along or near the line of the Colorado and Midland: Marion, Sunshine, and Spring Gulch Mines in Pitkin County, and Newcastle in Garfield County. Forty coke ovens were erected at the Marion Mine; when the coke proved inferior in quality, they were dismantled and the materials used to erect Belgium ovens at Cardiff. Coking coal was obtained from the Sunrise Mine in Spring Gulch, a few miles from Marion. The new ovens were rectangular in shape, approximately 4 feet wide, 20 feet long, with 7-foot-high arched roofs. They were designed to produce a grade of coke superior to that made at Crested Butte. The quality failed to meet expectations, but it was purchased by smelters at Leadville in sufficient quantities that the Colorado Coal and Iron Company curtailed plans for the further expansion of oven operations in Gunnison County.

The Grand River Coal and Coke Company fought with the Colorado Coal and Iron Company for control of coal lands in the valley of the Roaring Fork River. A three-year struggle between armed partisans of the two firms finally ended in 1889 with a division of the disputed territory. Properties north of Thompson Creek were assigned to the Grand River Company, and those to the south, including the area called Coal Basin, to the Colorado Coal and Iron Company. Two years later, John C. Osgood's Colorado Fuel Company purchased the Grand River Company, which then became part of the merger of the Colorado Fuel Company and the Colorado Coal and Iron Company, from which emerged in 1892 the CF&I.

Over a period of five years, the Grand River Coal and Coke Company operated 240 ovens to manufacture smelter coke. Forty ovens in operation at Marion in 1888 and 1889 accounted for 1,927 tons. The Cardiff plant produced 158,581 tons from January 1888 to October 31, 1892.

Cardiff was one of five coking plants acquired by the CF&I in 1892. That year El Moro, Starkville, and Sopris in Las Animas County, Crested Butte in Gunnison County, and Cardiff in Garfield County accounted for about four-fifths of the coke produced in Colorado. Coke that was not shipped to the Pueblo steelworks was sold to smelters of precious metal ores. As a manufacturer and distributor of coke, CF&I had no rival.

Its only in-state competitor, and a minor one at that, was the Victor Fuel Company. That enterprise had its origin with the Gray Creek Coal and Coking Company, organized by Delos Chappell, a prominent Trinidad businessman, in September 1887. The mine and ovens were six miles east of Trinidad. The Gray Creek Company merged in 1889 with the Colorado

Coke Company, another of Chappell's enterprises, to form the Victor Coal Company, which opened mines and erected ovens at Hastings, northwest of Trinidad. Chappell reorganized his fuel interests again in 1893 as the Victor Coal and Coke Company, with General Grenville Dodge and Morgan Jones, executives of the Denver, Texas, and Fort Worth Railroad Company, and Chappell being the firm's principal officers. They negotiated in 1899 the sale of the enterprise to John C. Osgood, who renamed it the Victor Fuel Company. Osgood planned to consolidate his new acquisition with CF&I, but before he found the opportunity to do so, he lost control of the Pueblo steel manufacturer. He retained control of Victor Fuel, and a successor firm, the Victor-American Fuel Company, until his death in 1926.

Other Colorado manufacturers of coke exerted little influence on the regional fuel market. The Pasadena Reduction Company, in 1886–1887, and the Grand View Mining and Smelting Company, from 1887 to 1896, made coke that was consumed by smelters in Dolores County. The Book Cliff Coal Company, operator of the Book Cliff Mine in Mesa County, manufactured small amounts of coke from 1892 to 1894 for domestic use in Grand Junction. Similarly, the Citizens Coal and Coke Company's thirty-six ovens in Denver produced coke for the domestic market in conjunction with the manufacture and distribution of coal gas to local customers.

La Plata County emerged as a center for the manufacture of coke in 1881, when John A. Porter, manager of the San Juan and New York Mining and Smelting Company, purchased the Green Smelter at Silverton and moved it to a site a mile south of Durango. The reduction furnace commenced operation in 1882, using locally made charcoal and coke. Local and eastern investors opened the San Juan Mine, a source of metallurgical coal, on the slope of the mountain above the smelter. Coal was lowered by means of a gravity-powered aerial tramway, 2,850 feet in length to the valley, where it was carbonized in beehive ovens. A double-track tram transported the coke a short distance to the furnaces.

In 1892 the Standard Smelting and Refining Company erected a reduction plant on the Animas River about a mile south of the San Juan Smelter. By then, the older plant consisted of six blast and ten reverbertory furnaces, plus twenty-eight beehive coke ovens. Only the San Juan plant survived the sharp contraction in the price of silver caused by the panic of 1893. The plant was leased in 1895 to the Omaha and Grant Smelting Company, of Denver. At that time, the coke-making plant of twenty-eight beehive ovens, if operated at capacity, produced about 11,000 tons of fuel a year. The Omaha and Grant Company was absorbed in 1899 by the American Smelting and Refining Company, the Smelter Trust, which purchased the plant in Durango and continued its operation for another three decades.

Four coke-making enterprises had a combined output of 557,305 tons in 1901. The CF&I led all rivals. Coking operations were located principally in Las Animas County, at El Moro, Starkville, and Sopris, with a combined total of 1,358 ovens. Its ovens represented 86 percent of the total in the state, and they produced 461,788 tons, or 83 percent of all the oven fuel manufactured that year. Most of the coke was consumed in blast furnaces at Pueblo to make pig iron.

At the time of its formation in 1892, CF&I acquired a small, antiquated metallurgical plant that president Osgood hoped to sell. When unable to do so, he decided to enlarge and modernize the steelworks in order to make it profitable. Because of the lingering effects of the depression triggered by the panic of 1893, reconstruction was delayed until 1899. Then, over the period of ten years, the plant was remodeled to turn out a full line of products, using state-of-the-art furnaces and rolling mills to maximize output and reduce costs. Expanded production required additional coke, as well as other raw materials, for making pig iron and steel.

Osgood pursued the twin goals of regional coal and coke monopolies. On his own account, he bought control of the Victor Coal Company, CF&I's principal competitor in the Colorado fuel market, in 1899, and opened what proved to be unsuccessful negotiations with the Union Pacific for its mining properties in Wyoming, Utah, and Colorado. Before Osgood had a chance to arrange for the CF&I to absorb the Victor properties, however, he lost control of the Pueblo steelmaker. The financial dislocation caused by the high cost of modernizing the plant forced Osgood to borrow heavily, and when he could not repay the loans, to surrender control of the company in 1903 to the principal lenders, George Jay Gould and John D. Rockefeller Sr. Thereafter, Osgood turned his attention to the Victor company, while the new management of the CF&I sought to complete the reconstruction program begun four years earlier.

The Colorado Fuel and Iron Company required additional sources of coke to support projected increases in metallurgical operations at Pueblo and to continue, at the same time, as a major supplier of fuel to smelters in the West. Officials had decided in 1899 to open additional mines and to erect new coking plants in southern and western Colorado. As the region's largest producer of pig iron, the Pueblo company wanted its own supplies of coke. Its blast furnaces, which ran without interruption for months, even years, at a time, required an ensured supply of fuel, for the price of coke largely determined the cost of reducing iron ore. Only by building and operating its own ovens could the enterprise exercise some control over the price of coke. Therefore, it followed the example of similar firms in the East by owning essential raw materials as a way of regulating the quantity, quality, and cost of the pig iron from which it made steel.

Already the largest manufacturer of coke in the West, the CF&I enlarged operations with new plants at Redstone in Pitkin County, and at Tabasco, Segundo, and Tercio in Las Animas County in the years 1899–1907. This construction program added 1,952 beehive ovens to operating units. Of these only Redstone was outside of the Purgatoire River Valley, in which Trinidad was the principal town and the administrative center for the CF&I's activities in southern Colorado.

The first of the new coke plants was erected at Redstone, in the Crystal River Valley south of Carbondale. Although partially completed in 1899, the ovens did not commence large-scale production until the following year. Coal for the ovens was mined initially at Placita, four miles south of Redstone on the line of the Crystal River Railroad, a wholly owned CF&I subsidiary. As soon as a branch, the so-called High Line, was built to Coal Basin, bituminous coal from that place was burned exclusively in the ovens. The original battery of 100 was later increased to 249 ovens. Redstone produced coke until June 1909, during which time it turned out 329,414 tons. Operations peaked in 1902–1903, when the output exceeded 65,000 tons. Most of the product was sold to smelters at Leadville.

The new community at Redstone was planned as a model industrial town. Osgood's decision to build a $2.5 million manor house there accounted for his special interest in Redstone and its residents. Company-owned dwellings for workers varied in size from three to five rooms, and were built to blend with the valley's natural beauty. Inside toilets, kitchens with running water, and electric lights—luxuries in mining communities elsewhere in Colorado—were common to all the homes in Redstone. The town featured a clubhouse and social center for employees and their dependents, a bandstand, public school, library, company store, and the Redstone Inn, a boardinghouse for single men. Osgood's palatial home was located above the town. Although he lost control of the CF&I in 1903, he retained ownership of his home at Redstone, and continued to reside there from time to time until his death in 1926.

Even before the ovens were completed at Redstone, some of the construction crews were transferred to the Trinidad area to build a new camp in Stock Canyon, a mile northwest of the Berwind Mine. A slope driven on a 5-foot seam supplied coal for a washery and 150 beehive ovens. As much as four-fifths of all coal mined was made into coke or shipped directly to Pueblo for the use of the steelworks.

The community, initially called Stock Canyon, was renamed Tabasco. By 1902 the camp had a population of 1,500, mostly Italian immigrants and their families. A total of 485 men were employed in the mine and at the ovens. Houses were grouped in two residential complexes approximately three-quarters of a mile apart, but equidistant from the mine and ovens. Four- and five-room cottages of varied design and color avoided the sense of sameness common to other camps in the district.

Two more large coking plants were built in the Purgatoire Valley. At a point eighteen miles west of Trinidad, contractors erected four batteries of beehive ovens at the new community of Segundo. For a time the largest facility of its kind in the West, the Segundo ovens were on the line of the Colorado and Wyoming Railway, a CF&I subsidiary, which ran from Jansen, a railroad switching yard on the west side of Trinidad, up the valley of the Purgatoire River to serve the corporation's mines and coking operations.

Segundo was located on the south side of the Purgatoire opposite the point where Smith's Canyon opened onto the river. Four miles up the canyon, the new Primero Mine produced the large volume of coal needed for the ovens. In this instance, as had happened twenty-five years earlier at El Moro, the coking units were placed some distance from the mine in order to gain easy access to water, which was consumed in large quantities at the ovens.

Fifty ovens were fired in August 1901. Another 750 commenced operation before the end of the year. All 800 were functioning by January 1903, turning out from 1,400 to 1,500 tons of coke daily. Approximately 450 men were employed, four-fifths of them at the ovens, and the remainder at two washeries.

Coal from the Primero Mine was run through a crusher, and then the washeries. The men who operated the latter facilities cleaned from 1,000 to 1,200 tons of crushed coal in a ten-hour day. The washed coal was

The Tabasco beehive ovens, northwest of Trinidad, Las Animas County, ca. 1915, were, by the standards of that day, a high-technology operation. The larry cars on top of the ovens were electric-powered, and a crew of six or seven men operated electric coke-pulling and car-loading machines. Coke-making ended at Tabasco in 1918. Courtesy, Bessemer Historical Society, Pueblo.

stored in bins until needed. When ovens were scheduled to be charged, larry cars were filled, driven across the tops of the ovens, stopping at each to drop the requisite amount of coal. The empty cars were gathered in strings by men driving mule teams and taken to a rope haulageway, on which they returned to the washeries. About half of the coke made at Segundo was shipped to the Pueblo works, and the balance was sold to smelters, including the Cananea Copper Company, Mexico; the United Verde Copper Company, Jerome, Arizona; the Copper Queen Company, Bisbee, Arizona; and the American Smelting and Refining Company for its plants in the western United States and Mexico.

Fourteen miles west of Segundo another coking plant was built at Tercio. Located in the foothills of the Sangre de Cristo Mountains, only four miles from Colorado's border with New Mexico, the new town was organized in 1901 as a community for workers who drove six openings ("drifts") there

The Tercio Mine and ovens (ca. 1915) were situated in the scenic foothills of the Sangre de Cristo Mountains, thirty-two miles west of Trinidad. Most of the coal produced from six drifts was consumed in the battery of 600 beehive ovens. Because of the declining demand for smelter fuel after 1900, only one-third to one-half of the ovens were in operation at any time, and coke-making ended in 1918. Courtesy, Bessemer Historical Society, Pueblo.

and in nearby Vega Gulch to obtain coal for two batteries of ovens. The coal was cleaned in two washeries, and converted into coke in 600 ovens. An additional 600 were planned but not built because of an unanticipated decline in demand for smelter coke.

A typical CF&I town, Tercio was built and controlled by the company. Prominent features were the clubhouse, company store, and physician's office, all staffed by the corporation's personnel. The fuel company advanced the money to build the school and churches. Houses varied in size from three to six rooms, and rented at the standard rate of $2 a room per month. The inhabitants were of many ethnic backgrounds, mainly Italian, with small numbers of Germans, Scots, Irish, and Swedes. "Little Italy," located between the tipple and the ovens, was composed of hastily built shacks to accommodate construction crews and their dependents. These were later replaced with company-built cottages. The predominantly Italian ethnic makeup of the community persisted for several years.

The new plants boosted the Colorado Fuel and Iron Company's output of coke. In 1905, Redstone, Tabasco, Segundo, and Tercio turned out a total of 586,503 tons, which exceeded the 451,550 tons credited to El Moro, Starkville, Sopris, Crested Butte, and Cardiff. Segundo alone accounted for 29.47 percent of the company's output.

By 1908 the Colorado Fuel and Iron Company, contrary to earlier expectations, produced more coke than it could market. This was attributed primarily to declining demand from smelters. Another factor was the failure of contractors to complete many of the improvements at the steelworks on schedule. As a result, blast furnaces and rolling mills did not reach planned levels of production that year. This prompted management to curtail coke production.

Retrenchment began in 1908. That year the CF&I closed the El Moro ovens. After the washer was destroyed by a fire in November 1907, the company declined to spend $60,000 to replace the structure and machinery. The Redstone ovens were abandoned in 1910, at which time the mine at Coal Basin was closed. The market for coke in the mountain district had declined to the point where the Redstone plant could not be operated at a profit. That was also true of Cardiff, where coking operations ended in 1910. Only a small amount of coke, an average of about 60 tons a year, was made from 1911 to 1914.

The remainder of CF&I's beehive-oven plants remained in operation until the company inaugurated by-product oven operations at the steelworks in 1918. That year plants at Starkville, Sopris, Tercio, Crested Butte, and Tabasco were abandoned. Some ovens at Segundo remained in operation to supply what remained of the market for smelter fuel, and to provide a reserve, if needed, for the steelworks. For the next eleven years, coke production at Segundo averaged 68,715 tons, and a maximum of 163 ovens were active at any one time. A fire wrecked most of the plant in 1929, after which the corporation ceased making coke in beehive ovens.

Colorado Fuel and Iron and other manufacturers of coke continued the use of beehive ovens long after they had been abandoned by metallurgical companies in the East and Midwest. Beehives remained popular in the West because of their adaptability to local conditions. They were ideally suited to an inherently unstable industry. The reduction of precious metals, as well as iron ore, fluctuated frequently from month to month and from year to year, as did the need for coke as furnace fuel. Beehives could be built and maintained at relatively little cost. They were easily shut down and restarted on short notice. Output could be readily adjusted to changing market conditions. When demand increased, crushed mine-run coal could be diverted to the ovens. If coke sales slumped, coal bypassed the ovens and was offered to residential and industrial consumers.

Furthermore, operators preferred beehives because it was cheaper to transport coke than coal. A third or more of the weight of coal was lost in the production of coke. Ovens were erected at mine sites, always with railroad connections, and the product was distributed to a number of customers throughout a large area. Beehives were simple but effective tools for making coke, and the fact that they were extremely wasteful apparently bothered no one as long as smelters consumed large amounts of coke to reduce precious-metal ores.

By 1910, however, the smelter market had largely disappeared. Smelting, a method of separating precious metals from ores by means of heat, was highly efficient, but costly because of the expense of building and operating reduction-furnace plants. As long as Colorado mines produced rich ores that contained hundreds of dollars' worth of gold or silver per ton, the cost of smelting seemed negligible. When rich ores gave way to low grades, as happened inevitably in all mines, owners and operators turned to the more efficient and affordable chemical mills that used solutions of cyanide, chloride, or bromide to extract small amounts of precious metals from large amounts of ores. By 1910, the shift from smelters to mills was an established trend that forced manufacturers of coke to cut output.

Coke-makers responded in different ways. The Union Coal and Coke Company, operator of ovens at Gulch, in Pitkin County, sold its properties to the Rocky Mountain Fuel Company in 1902, and the new owner abandoned the ovens, preferring to focus on mining and marketing coal. The Victor Fuel Company built an additional eighty ovens at Delagua, near Hastings, in 1904. Five years later the firm merged with the American Fuel Company to form the Victor-American Fuel Company. Its thirteen mines in Colorado and New Mexico had a combined output of 12,000 tons of coal a day, and three coke plants, with a total of 445 ovens, had a combined capacity of about 500 tons a day. Oven operations ended at Delagua in 1910 and declined thereafter at Hastings and Gray Creek. There was a brief revival of output from 1915 to 1917, after which coke-making was abandoned because copper companies in the Southwest adopted new methods of ore reduction. Victor-American mined and marketed coal for another three decades.

The American Smelting and Refining Company (ASARCO), Colorado's only other operator of beehive ovens, enlarged operations as many enterprises scaled back the output of coke. In 1899, ASARCO acquired the small coke-oven plant at the San Juan Smelter in Durango. Over a period of three decades, from twenty to thirty-four ovens were active at any given time, turning out as much as 15,000 tons of fuel a year. The smelter and the ovens ceased operation in November 1930, during the early months of the Great Depression. By then the company's major coke-making operations had shifted to Las Animas County.

The company decided, probably in 1906, to own and control the source of coke for its vast smelting operations in the United States and Mexico. For that purpose it created a subsidiary, the Carbon Coal and Coke Company, to open mines and organize a company town, Cokedale, north of the Purgatoire River at the entrance to Reilly Canyon, seven miles west of Trinidad. The physical plant consisted of two mines, a washery, and a double battery of 380 beehive ovens, houses, and service facilities for a community of several hundred people.

Cokedale was a model town. One hundred fifty houses, ranging in size from three to eight rooms, were constructed of concrete blocks on foundations of concrete-lined cellars. Many of the houses had bathrooms with tubs and toilets, and all were wired for electric lights. Other structures in the community included a store, clubhouse, bathhouse, hotel, company office, power plant, and two-story elementary school.

The Reilly Canyon coking plant seldom operated at capacity. Output remained consistently below 200,000 tons a year until the 1920s, then dipped sharply during the recession of 1921–1922, and recovered to previous levels until the general economic collapse in 1929. Output dropped below 65,000 tons a year in the 1930s. A dramatic recovery during the early months of World War II was nipped in 1942 by the federal government's ban on gold mining for the duration of the conflict. The opportunity for postwar recovery vanished when the Boncarbon Mine, the source of coal for the coke ovens, closed in May 1947.

ASARCO was, with one exception, the last manufacturer of beehive-oven coke in Colorado. Mid-Continent Resources Inc., owner and operator of mining properties near Redstone, refurbished forty-seven of the old ovens to make special-order coke for customers in California in the years

1959 to 1961. After that brief revival of activity, the plant was again abandoned.

Beehive-oven coke-making was, in fact, an anachronism long before its final production in Colorado. When the Pueblo steelworks became the principal consumer of coke in the state, it was no longer cost-effective for the Colorado Fuel and Iron Company to operate beehive ovens at mine sites. By relocating production to the point of consumption the corporation could eliminate waste and lower costs by capturing and marketing the by-products that formerly escaped from the ovens to pollute the atmosphere. Retorts produced superior grades of coke for the iron-ore reduction furnaces, and provided additional income from the sale of a variety of products derived from the gases given off by coal during the carbonizing process.

By-product coke-making had been widely used in Europe long before it was accepted on a large scale in America. The first retorts in the United States, twelve Semet-Solvay ovens, were erected at Syracuse, New York, in 1893. The distillation of coal yielded gas as the primary product, plus ammonia, benzol, and tar from which a variety of chemical products were manufactured. Coke was a secondary product. The first retorts at an American steelworks were Otto-Hoffman ovens installed at Johnstown, Pennsylvania. Rothberg ovens were built at the Lackawanna Steel Company in Buffalo, New York. The first Koppers regenerative units were erected at the Illinois Steel Company's Joliet plant in 1907. When the United States Steel Corporation added Koppers ovens at Gary, Indiana, they became the standard for the American steel industry.

The Pueblo steelmaker considered the construction of by-product ovens as early as 1899, but put off a decision for another sixteen years. In 1915, CF&I president Jesse F. Welborn signed a contract for the first ovens of any kind to be built west of St. Louis. Construction commenced the following year, and the ovens were fired in July 1918. One hundred twenty regenerative ovens, arranged in two batteries, with a washery, jigs, and equipment for capturing and processing by-products, cost approximately $5 million.

The ovens, each a closed chamber surrounded by a flue, were arranged in batteries of sixty. Each oven measured approximately 40 feet in length, 10 feet in height, and 18 inches in width. Larry cars, transporting coal that had been washed and crushed to a uniform size, passed along rails above the batteries, dropping a charge into the top of each oven. That coal was leveled by an electrically driven bar inserted through a small opening in the oven door, after which burning gases in flues mounted in the oven walls heated the coal to temperatures of about 2,000 degrees Fahrenheit to drive off gases and tar. The process required from sixteen to twenty-four hours, approximately half the time of beehive ovens. The length of the burn was dictated by the desired quality of coke. After completion of the burn, the oven doors were removed and a mechanical ram shoved the contents into a quenching car for cooling. That operation was spectacular, for water, on striking the hot coke, produced a huge, billowing cloud of steam. After the coke had cooled, it was screened and classified as either furnace or boiler fuel. Furnace fuel was stored until required for reduction of iron ore in the blast furnaces. Boiler fuel, called breeze or fine coke, was unsuited for the furnaces. It was used as bedding for the ingot-soaking (heating) pits in the rolling mills. Most of the coke was consumed in the Pueblo plant, but some was sold to foundries and to regional enterprises that made sugar from beets.

By 1923, the by-product oven department consumed approximately 30 percent of the coal produced at the CF&I's mines. Fuel for the ovens, about 3,000 tons per day, came mainly from mines near Trinidad and Crested Butte. At that time approximately 25 percent of the company's coal output was sold as boiler fuel, an equal amount for space heating in homes and businesses, 5 percent to regional public utilities, and the rest, except what went to the coke ovens, was consumed as boiler fuel at the steelworks.

The Pueblo plant produced about 1,700 tons of coke daily, plus numerous by-products. The 26 million cubic feet of gas generated by carbonizing coal every twenty-four hours was drawn off, cooled, and subjected to a series of treatments to extract valuable chemicals. Each ton of coal charged in the ovens produced, in addition to coke, ammonia sulfate, benzol, and tar. The large volume of gas remaining after by-products had been removed was used mainly to heat the ovens. Any surplus was consumed in the plant's blast furnaces and the reheating furnaces in the rolling mills.

Ammonia sulfate was produced by passing coke-oven gas through a saturator containing a dilute solution of sulfuric acid. Ammonia in the gas combined with the acid, forming crystals that collected in the bottom of the saturator. These were removed, washed, dried, mixed with gypsum to prevent caking, bagged in 100-pound units, and marketed as agricultural fertilizer. Benzol was extracted from gases by scrubbers. It was sold either to petroleum refineries to be blended with gasoline as motor fuel or to manufacturers as a solvent for making paints. Tar, the residue of the gas-cleansing process, was consumed at that time as fuel in the open-hearth furnaces where pig iron was transformed into steel.

In early 1929, when CF&I was still enjoying prosperity, the board of directors authorized the construction of the first of two additional batteries of Koppers ovens. By that date, Batteries A and B, in operation for a dozen years, had become obsolescent, if not obsolete, and their limited capacities were stretched beyond safe limits in an effort to meet the needs of the Pueblo plant for blast-furnace coke. Because of the general depression that dampened the company's operations before the close of the year, only one battery was completed at that time; the other one was indefinitely delayed.

Battery D (C was in abeyance), composed of thirty-one ovens, added 380 tons a day to the plant's capacity. The new ovens, operational in 1930, were larger and more efficient that the earlier models. In conjunction with the ovens, the company built a smokestack more than 250 feet in height. One hundred feet taller than any other chimney at the plant, it was a landmark for travelers throughout the area.

With the new battery, the coke plant, when operated at or near capacity, consumed approximately 1 million tons of coal a year to produce 700,000 tons of coke. There was a proportionately large output of by-products, which CF&I decide to fabricate and to market as its own brand of industrial and household chemicals. Commencing in 1929, the corporation's chemists distilled coal tar to make creosote on a large scale for sale to enterprises for weatherproofing wood products. It was widely used for treating railroad ties. In addition, roofing pitch, derived from tar, was sold throughout the area. Any tar that was not converted into marketable products was consumed, as in the past, by the open-hearth department.

3

Lamp Oil to Motor Fuel 1860–1930

Petroleum was discovered in Colorado in 1860. Crude oil skimmed from the surface of Four Mile Creek, northeast of Cañon City, was refined in simple stills and sold as lamp oil and lubricant. The persistence of the seepage suggested that a large underground pool could be tapped by sinking wells on the site. Attempts to drill for oil were not successful in the 1860s and 1870s. In 1881, a driller of a water well near Florence, several miles east of the seepage, encountered oil. News of that discovery led to exploration and development of the Florence field, for many years the principal source of the petroleum products consumed in the Rocky Mountain region.

Gabriel Bowen, a hunter-trapper, was credited with discovering the oil springs in September 1860. He reported that the oil flowed at a rate of about 5 gallons an hour, and in appearance and smell it resembled "coal oil," the lamp fuel distilled from coal and widely used in eastern states. Bowen staked a claim and prepared to collect the oil for resale.

News of the discovery quickly spread to other parts of the territory. Samples of Cañon City oil were put on public display in Denver. The *Rocky Mountain News* announced that Bowen planned to sink a well in order to increase the output of oil for commercial purposes. For some time, however, the only source of oil was three or four seepages, one of which

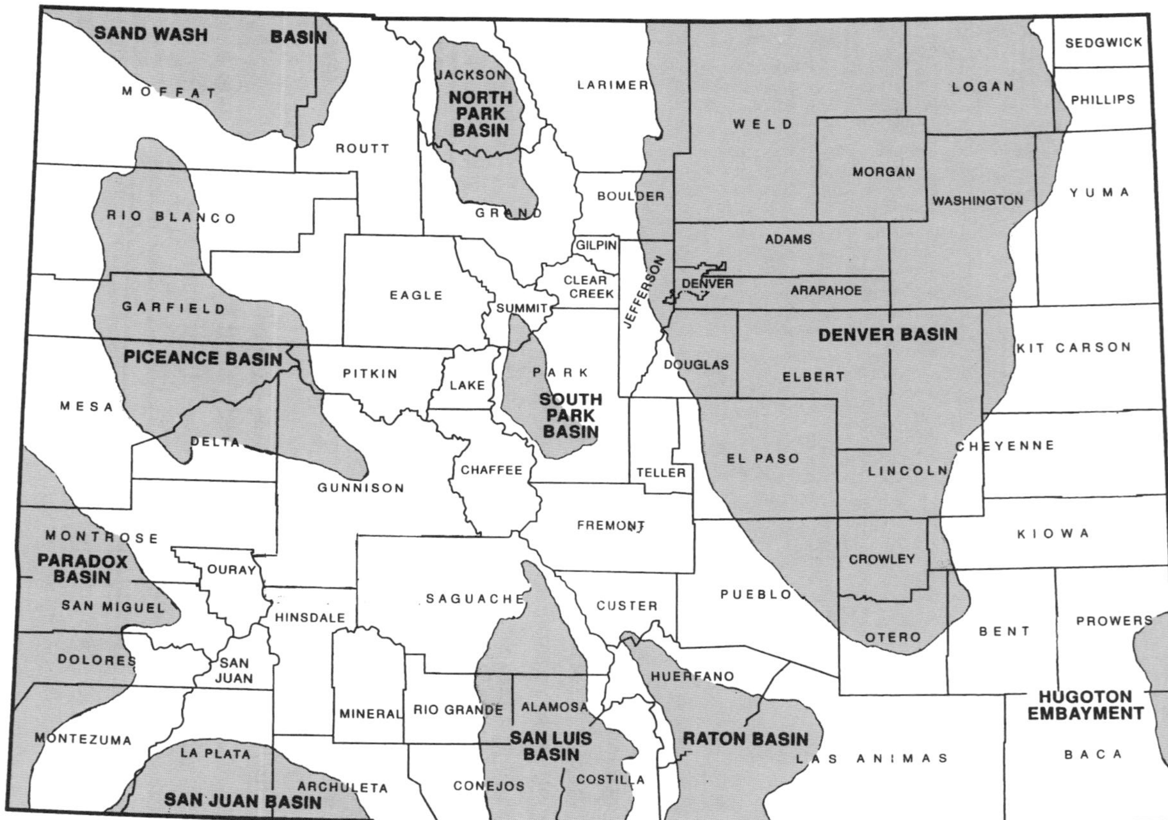

Map 2. Oil and Gas Regions of Colorado. Adapted from Colorado Geological Survey, *Colorado Mineral and Mineral Fuel Activity, 1997, Information Series 44* (1998).

yielded an almost pure product. In 1862, Bowen sold the tract to Alexander Morrison Cassiday, who planned to sink wells.

Cassiday was born in Wayne County, Ohio, in 1827, to parents who sought better farming land, first in Richmond County, Ohio, and later in Iowa. As a young man, he studied law and pursued that profession for a number of years in Oskaloosa. He may have served a single term in the Iowa state legislature before joining the gold rush to the Pikes Peak region in 1859. He practiced law in Denver before moving, in 1862, to Cañon City. Aware of Bowen's discovery and of his efforts to market petroleum products, Cassiday dreamed of making a big oil strike by tapping what he believed to be a huge pool of oil that fed the seepages. He erected a still to process oil skimmed from the surface of the creek. Refined oil was transported by bull teams to Denver and other communities.

Cassiday sank at least three, perhaps as many as six, wells on the tract along Four Mile Creek. Some were dug by hand; others were drilled with crude spring-pole equipment that he probably assembled on the site. Contrary to expectations, the wells did not increase the flow of oil, which remained, according to Cassiday's calculations, constant at about one barrel (42 gallons) a day. Over a period of nearly three years, Cassiday refined and sold several thousand gallons of oil. Lamp oil sold in Denver, Pueblo, and Santa Fe for $1.25 to $2.85 per gallon. Lubricating oil brought about 65 cents a gallon. Scarcity caused by the Indian war of 1864–1865 briefly boosted the price of illuminating oil in Denver to $5 a gallon.

In 1864 Cassiday sold his petroleum business to Benjamin Roop, who may have been his father-in-law. The operation of the wells and refinery came under the direction of James Murphy. In 1869, a member of F. V. Hayden's geological survey team visited the site. It was, according to Persifor Frazer, very primitive. Two wood-casing wells, each about 300 feet deep, had a combined output of about 4,000 gallons a year. Murphy planned to tap a larger supply of oil by drilling deeper. There is no record that he did so. By the early 1870s, the enterprise was inactive. The construction of the Kansas Pacific and Denver Pacific railroads to Denver opened the Colorado market to eastern refineries, who sold lamp fuel and lubricants at a fraction of the cost of Cañon City oil.

After disposing of his oil property, Cassiday turned his attention to other enterprises. He became a partner in the Grand Canyon Coal Company, owner of a mine south of Florence, near the town of Coal Creek. With capitalists from Cleveland, Ohio, he acquired gold mines near

The United Oil Company's refinery at Florence, Colorado, as it appeared in the 1890s. Built in 1885, it remained in operation until 1936. Prior to 1900, this plant produced most of the lamp oil and other petroleum products consumed in the eight states and territories of the intermountain West. Continental Oil Company bought the refinery in 1916. Courtesy, Cañon City Public Library, Local History Center.

Rollinsville, in Boulder County. That venture was a failure. He recruited Boston capital to underwrite exploration in the vicinity of Cañon City and Florence for the oil pool that he believed fed seepages on Four Mile Creek. At a depth of 600 to 700 feet he expected to find more than enough crude oil to supply the whole of the Rocky Mountain region.

In an effort to tap the oil pool, Cassiday turned to Ira and Isaac Canfield, who owned a set of drilling tools. A partnership of Cassiday, the Canfields, and Dr. M. H. Slater, formerly of Denver, sank a well near Cañon City commencing in March 1878. When they did not find oil, Isaac Canfield moved the derrick and machinery to the property of the Grand Canyon Coal Company, near Coal Creek, to drill for water, which was needed to support mining operations. In December 1880, drillers began pounding pipe into the ground. On January 21, 1881, they encountered evidence of oil at a depth of about 1,260 feet. Operations were halted for fear of a "blowout," because natural gas spewed from the well. The next day, the crew used a baler to extract oil from the hole, but only about ten barrels were brought to the surface. Canfield, convinced that the drill had penetrated a pocket rather than a pool of oil, resumed sinking to greater depth. Shortly after restarting operations, the manila-rope cable broke, allowing the tools to fall to the bottom of the hole. Efforts to retrieve them were unsuccessful and the well was eventually abandoned.

The Canfield well, while not a commercial success, created great excitement. People rushed to Florence by train, and on to Coal Creek on foot or by carriage to see the well. Curious sightseers carried small tins and bottles that they filled with the dark oil. Land values in the vicinity of the well skyrocketed, and fortune seekers sought leases from landowners in the hope of sinking wells of their own.

Among the first to take advantage of the excitement was a group of local people organized as the Land Investment, Coal and Oil Mining Company. The firm's president, David G. Peabody, a native of Vermont, had engaged in the mercantile business in Denver for a dozen years before moving to Cañon City in the early 1870s. Peabody and associates secured leases on what looked like promising land. They started a well in November 1882, and on April 7, 1883, encountered oil at a depth of 1,205 feet. After a slow start, the well became a large producer.

This success prompted the formation of another venture. A. M. Cassiday; his son, D. R. Cassiday; Isaac Canfield; and others organized the Arkansas Valley Oil Company in September 1883. Two years later, that corporation was reorganized as the Arkansas Valley Land and Oil Company. New investors from Ohio brought Daniel P. Eells, a Cleveland banker, and his associate, John Coon, to Florence. Oil from several wells completed in 1883 and 1884 was processed in a new refinery. The 200-barrel-a-day plant, erected in 1885, was quickly enlarged to handle the crude-oil output of other enterprises. The company sold illuminating and lubricating oil locally, as well as in Denver, Pueblo, Aspen, and Leadville.

Meanwhile, the Land Investment Coal and Oil Mining Company, after sinking two wells, experienced a diminished cash flow that prevented payments on mortgages taken out to acquire land. D. G. Peabody bought the indebtedness and put the land into a new venture, the Colorado Oil Company, which he organized with Jacob Wallace, a recent migrant to Florence from the East. Augustus R. Gumaer and S. A. Josephi joined Peabody and Wallace in the venture. In two years, the corporation completed sixteen producing wells.

A successful New York City entrepreneur, Wallace had decided in 1884 to make a tour of the West in search of business opportunities. The train on which he traveled across Colorado stopped in Florence for coal and water. Wallace walked about the small community, noticed the infant oil industry, and concluded that it had great potential. When the train departed, Wallace remained behind to talk to Peabody and others who were putting down wells.

By 1887, Wallace concluded that the principal enterprises at Florence should be consolidated into a single corporation. In July, he persuaded the Arkansas Valley Company to join the Colorado Oil Company in a new concern, the Colorado Oil Trust, headed by Nathaniel P. Hill. That corporation had an authorized capital of $3 million. Former U.S. senator Hill, who had erected the state's first successful smelter at Black Hawk, was president of the firm, but operations were in the hands of Wallace as vice-president and general manager. D. P. Eells was another vice-president, and S. F. Rathvon was secretary and treasurer. Other directors were Isaac E. Blake, John Coon, and S. A. Josephi. Renamed the United Oil Company, it became the field's leading producer and refiner of crude oil. By terms of an agreement in 1888, its entire output of petroleum products was marketed by the Continental Oil Company.

That enterprise had its origin in Ogden, Utah, where in 1875 Isaac Blake persuaded a group of local businessmen to form the Continental Oil and Transportation Company, which purchased petroleum products, mostly kerosene, in the East for distribution throughout the Rocky Mountain states and territories. A Denver office, opened in 1876, was under the management of E. R. Burton. Initially, kerosene was sold in 5-gallon metal cans and was widely consumed as lamp oil. Burton introduced bulk stations from which oil was peddled by tank wagons to communities throughout the region, enabling customers to buy according to their needs, whether large or small.

Continental's principal competitor was the Consolidated Tank Lines Company, which operated its own bulk stations and wagons. In 1882, Standard Oil interests gained control of Consolidated, and immediately slashed the price of kerosene in an effort to enlarge its share of the market. Blake, faced with the prospect of heavy losses, agreed to a merger of the two firms in 1884. The Continental Oil Company was incorporated in Iowa to market petroleum products throughout the Rocky Mountain region, the Pacific Coast states, Alaska, and a portion of Canada. Blake was president, but control of the enterprise was in the hands of John D. Rockefeller and associates. Continental was part of Standard Oil's national monopoly for marketing petroleum products. Similar enterprises existed in other parts of the country. Under the watchful eyes of the Standard Oil trustees, affiliates were assigned specific territories and encouraged to overcome all competition in servicing their markets.

Another producer and refiner was the Florence Oil and Refining Company. Organized in January 1887 by Appleton H. Danforth, Augustus R. Gumaer, and William E. Johnson, that enterprise was the United company's leading rival. Isaac Canfield was for many years vice-president and general manager of the firm. Danforth had formerly served as general manager of the Colorado Coal and Iron Company, and Gumaer and Johnson were local entrepreneurs who were active in ranching, banking, and mercantile activities. Canfield, with his father, owned the Triumph Oil Company, which sold crude oil to the Florence company's refinery. All petroleum products were distributed by the Standard Oil monopoly.

In December 1889 the Florence Oil and Refining Company canceled its agreement with the Continental Oil Company in order to market its own products. In the resulting "oil war," Continental reduced the price of kerosene from 15 to 5 cents a gallon in Denver and other cities in Colorado. The Florence Oil and Refining Company matched the reductions for a time, but did not have the resources to operate indefinitely without profits from sales. After two years, the firm capitulated. It temporarily closed its refinery, restricting operations to the production of crude oil that was sold to an independent corporation, the Rocky Mountain Oil Company. In 1895, the Florence Oil and Refining Company opened a new refinery whose products were marketed exclusively by the Continental Oil Company.

The Rocky Mountain Oil Company was organized in 1890 by several men who had been active in the affairs of the United Oil Company. Led by Dan P. Eells, a group of Cleveland residents broke from the older firm to establish a new enterprise to produce oil and market petroleum products free of the constraints imposed by other corporations. John Coon, as superintendent, was in charge of all operations. Crude oil was processed in a refinery erected at Overton, six miles north of Pueblo. Via a 4-inch pipeline oil flowed twenty-eight miles from the field to the plant, which had a capacity of 1,000 barrels a day. The company marketed illuminating oil, lubricants, and paraffin, refusing the services offered by the Continental Oil Company.

The Rocky Mountain Oil Company was a challenge that the Standard Oil Trust could not ignore. Continental Oil slashed prices for kerosene, and the newcomer followed suit. The price of kerosene fell to as little as 5 cents a gallon in Pueblo. At one time, Continental allegedly gave its regular customers in some communities lamp oil without charge. After eighteen months, the Rocky Mountain company, its finances in total disarray, capitulated. It closed the refinery at Overton and sent crude oil to the United Oil Company for processing.

Bankruptcy ended any hope of reviving operations at Overton. In 1898, Eells, acting in behalf of a majority of the stockholders, purchased at tax auction the Rocky Mountain Oil Company's land, buildings, machinery, and oil leases. He reorganized the enterprises under the same name, but its activities were limited to the production and sale of crude oil. The pipeline was sold to a California firm. The refinery was dismantled and a portion of it was eventually relocated to an oil field near Boulder. In 1903, the Rocky Mountain Oil Company again ceased operations. The United Oil Company purchased the corporation's assets, including 1,600 acres of land, numerous wells, and supporting facilities.

The Florence field, when fully developed, covered an area of about fourteen square miles. It included the town of Florence, where, in the early days, most of the Florence Oil and Refining Company's wells were located. The northern limit, although not well defined, was the Arkansas River. Wells were drilled for a distance of about four miles to the south, with a heavy concentration around Coal Creek, and nearby towns on Oak Creek—Williamsburg, Rockvale, and Chandler. The field's average east-west width was about three miles.

For almost two decades, Florence's refineries were the exclusive source of petroleum products for the states and territories of the intermountain West. Kerosene was extracted by heating crude oil in stills. Lighter fractions, including gasoline, for which there was no market at the time, were wasted in order to produce oil for lamps. About one-third of the crude oil was consumed in making lamp oil and lubricants. The residual oil and tar were sold as boiler fuel, or burned under the refinery stills. In the 1890s local and regional ore-reduction mills purchased the residual oil for roasting ores to remove sulfur, and used tar as boiler fuel. The fuel oil sold for 50 cents a barrel, or slightly more than 1 cent a gallon. While oil, in this instance, competed with locally produced coal, the supply at that time was too small to have a significant impact on the fuel market.

For about twenty years Florence's refineries produced ample oil products to meet the region's needs. From less than 100,000 barrels in 1887, the first year for which reliable statistics were collected, the output rose to a peak of 824,000 barrels in 1892. After that year, the annual production fell gradually to 317,000 in 1900. By that date, the demand for petroleum products in the intermountain West exceeded the capacity of the refineries at Florence. The field's output surpassed 200,000 barrels of crude oil for the last time in 1915. However, oil production continued at Florence for several decades, and refining operations were not shut down until 1936.

At the turn of the century, it was apparent to experienced oilmen that the Florence field was past its peak production; it could not supply a growing demand for petroleum products in the Rocky Mountain region. For that reason, there was a widespread search throughout Colorado for new petroleum resources. In spite of intense exploration, only two fields were discovered in the years 1901–1920. A small one at Boulder, northwest of Denver, was fully developed. The other, at Rangely in Rio Blanco County, was not developed commercially at that time because there were no transportation facilities to move oil to distant markets.

The Boulder field was discovered and developed by men who had been active in the oil industry at Florence. The initial well was put down by a crew of drillers supervised by Isaac Canfield, who had discovered the Fremont County field. With support of Denver and Colorado Springs investors, Canfield organized the Boulder Oil Company, a subsidiary of the Consolidated Oil Company, to underwrite the cost of exploration and development. The president of the parent firm, C. H. White, was chief executive of the El Paso National Bank in Colorado Springs. Charles Page, who had lived for some time in the Cripple Creek district, was secretary and treasurer. As vice-president and general manager of both enterprises, Canfield secured leases on the Neil D. McKenzie farm, near Boulder, and the Dodd ranch at Niwot. Canfield said he was drawn to Boulder because the topography resembled that of Florence. He was convinced that the land along the foothills from Boulder to Fort Collins was underlaid with a large pool of oil.

Canfield's choice of the Boulder area may have been influenced by more than topography. Wells put down in the vicinity of the town as early as 1891 had encountered oil, but not in commercial quantities. Geologists who studied the upturned geologic strata in the foothills insisted that

Derricks standing in the Boulder oil field (ca. 1905). Isaac Canfield, who completed the first oil well at Florence, put down the McKenzie No. 1 northeast of Boulder in 1901. It penetrated oil, setting off a stampede to secure oil leases. The well collapsed, and was followed in 1902 by the commercially successful McKenzie No. 2. The field was small in size and output, but produced a superior grade of crude oil. Courtesy, Denver Public Library, Western History Department.

crude petroleum existed under much of northeastern Colorado. They predicted that a potentially rich field would be found from Denver northward into Wyoming, and eastward from the foothills for many miles. However, at some distance from the foothills drills might have to penetrate to depths of 4,000 feet or more to find the elusive prize. Because of a growing scarcity of oil in Colorado, many of the same geologists agreed that an exploratory well should be drilled to find the deep oil strata. As a few were quick to admit, depths of 4,000 feet or more were probably beyond the reach of the percussion tools currently used to sink wells.

It is also likely that a famous explorer-geologist influenced Canfield's selection of Boulder as a site for an exploratory well. Ferdinand Vandiveer

Hayden, one of the founders of the U.S. Geological Survey, completed a scientific reconnaissance of Colorado, including the area northwest of Denver, in the 1870s. He published a map that identified a narrow area along the eastern foothills of the Rocky Mountains from Morrison and Golden to the Colorado-Wyoming border, and beyond, as having geological formations that suggested underlying oil.

Canfield's test commenced in 1901, when a drilling crew spudded a well on the McKenzie farm, near the Colorado and Southern Railway tracks between Boulder and Six Mile Reservoir. Another crew pounded pipe into the ground at Niwot. Canfield selected both sites with the aid of a "bobber," a witch-hazel branch similar to that used by locators ("dowsers") of water wells. On August 7, Canfield reported that the McKenzie, at a depth of 1,700 feet, had encountered a mixture of water and oil. The drillers continued to greater depth, while C. H. White ordered fireworks that were set off on the evening of the thirteenth to celebrate the discovery of oil. On August 22, the drill penetrated oil at a depth of 2,500 feet. Samples were removed until the rope cable broke, sending the baler to the bottom of the well. It had to be retrieved before pumps could be attached to the well. Efforts were restricted to daylight hours, for open-flame torches used to light the area at night would have ignited the natural gas that spewed from the hole. Canfield admitted defeat on September 14 after the wall of the uncased well collapsed. An attempt to move the derrick intact was unsuccessful. It was dismantled and reassembled elsewhere on the McKenzie property to sink McKenzie No. 2.

The high-grade oil taken from the McKenzie No. 1 created great excitement. Numerous enterprises were organized to drill for oil, promising investors large returns on purchases of stocks. By January 1902, companies were prepared to put down wells at many places in Colorado. Exclusive of the Florence field, 48 wells were completed by the following month. At the same time, another 101 derricks were under construction, suggesting an accelerating pace of exploration and development. The focus of activity was Boulder, but new wells were sunk at De Beque, in Mesa County; Pagosa Springs, in Archuleta County; Fort Collins, in Larimer County; and Morrison, in Jefferson County. Exploration also occurred in wells in Arapahoe, Elbert, El Paso, La Plata, Logan, Pueblo, and Saguache Counties.

The success that crowned Canfield's efforts at Boulder was not repeated at Niwot. The well on the Dodd ranch reached 2,500 feet without penetrating oil. Concluding that the depth of the mineral probably increased at points north of Boulder, Canfield continued the search. In early November 1901, he finally abandoned the effort at 3,600 feet. At the urging of C. H. White and Charles Page, he redirected his attention, at least for a time, to the vicinity of Boulder. As the Consolidated Oil Company's general manager, Canfield had charge of putting down a well on the William Arnold ranch, one mile northeast of the McKenzie farm. The Arnold well, completed on January 6, 1902, was the Boulder field's first large producer of oil. The McKenzie No. 2 struck oil four days later. Two commercial wells, in quick succession, convinced even die-hard skeptics that a major new source of petroleum had been discovered.

The Consolidated Oil Company was the toast of the town and the state. Honors were heaped upon White and Page for having taken on a heavy financial burden in backing Canfield's quest for oil at Boulder. Their McKenzie No. 1, although credited with the discovery of oil, earned no profits. The well on the Dodd ranch at Niwot was a costly duster. The Arnold and McKenzie No. 2 allowed them to recoup their losses.

Both men devoted much of their time and money to exploration for oil. White became one of the most active promoters in the Boulder field; his Consolidated Oil Company owned or controlled numerous enterprises, including the Headlight, Footlight, Alamo, Blue Jacket, Crystal, Interstate, and Lafayette oil companies. Page sold his Boulder oil interests in 1902 and eventually relocated to Tulsa where, by the 1920s, he owned large oil and gas interests, two newspapers, and was acknowledged as the city's leading philanthropist.

Page left Boulder at a time when the influx of investors imposed dramatic changes on the quiet, once-secluded university town. It became, at least for a time, a bustling mecca of gamblers, fortune seekers, and speculators who vied with each other in a frenzy to create oil companies and to sell stocks to the public. Oil securities were bought and sold, and resold again and again. Most of the enterprises were incorporated in Wyoming, where, unlike Colorado, stocks could be issued in denominations of less than a dollar a share. Stocks that sold for 1–10 cents a share had great appeal. Investors probably gained little in the way of dividends, but promoters, many of whom peddled securities of concerns that existed only on paper, reaped large profits. Most of the new ventures proposed to search for oil in the Boulder field; others planned similar activities elsewhere in Colorado and in Wyoming.

In the public's view, Canfield, and those who followed, validated Ferdinand V. Hayden's prediction of Boulder's potential for oil production. While at Boulder, Hayden had been a guest of Ross Howard, a rancher. A popular story related that Hayden, overwhelmed by Howard's gracious hospitality, walked one morning to a point in the corral, drove a stake in the ground, and informed his host that it marked the location of an immense underground lake of oil. Howard, who was cautioned to hold on to the land because of its great value, put down one well to a depth of 260 feet without finding petroleum. Following the completion of the McKenzie and Arnold wells, many people spent an inordinate amount of time looking for Hayden's stake.

Several enterprises claimed to have found the exact location of the stake. Predictably, they offered the public an opportunity to share in their good fortune through the purchase of stock. The Amber Oil Company was the first to announce that it was putting down a well on the spot allegedly marked by Hayden. The outcome was a disappointment. In July 1902, the Boulder Mining, Oil and Gas Company identified another site as the location of the stake. It was in the midst of producing wells—the Bradford on the north, the Bingham on the east, and the Boulder Petroleum on the west. Therefore, it was not surprising that the corporation's three wells, Hayden Nos. 1 through 3, were large producers of oil.

By 1903, Canfield concluded that the Boulder field was only a small pool, the output of which would have little impact on the region's needs for petroleum products. Consequently, he left Colorado in search of new fields. He and his son, Carl, became itinerant drillers, moving from place to place, stopping wherever there was need for their services. They were active for a time near Calgary, Alberta, where they drilled forty-one wells, thirty-eight of which produced oil. They worked in Kansas, New Mexico, and Texas, and from 1912 to 1920 near Muskogee, Oklahoma. Eventually Canfield returned to what he called his permanent home, the Buckhorn Ranch in the Plateau Valley, near Collbran in western Colorado. After his wife died, he lived with a daughter in Grand Junction. He died there on October 15, 1924.

The Boulder field, as Canfield anticipated, did not live up to the claims of stock promoters. Its output of oil never equaled as much as one-fourth that of Florence. From a modest 11,800 barrels in 1902, production rose to 36,722 barrels in 1903, then fell gradually to 10,502 barrels in 1905. At that time, wells at Florence pumped about 400,000 barrels a year. The quality of the oil compensated, at least in some measure, for the small quantity of the output. High-grade with a paraffin base, it was, according to the U.S. Geological Survey, superior to any found as of that date west of the Mississippi River. Unfortunately, development was haphazard, without concern for prolonging production. Wells were abandoned, and gas, in the absence of regulations, was allowed to escape from unplugged holes. The loss of pressure in the underground reservoir led inevitably to a reduction in output. However, early predictions that the field was on the verge of total demise were premature.

Exploration to the north of the initial area of development boosted output to a record 48,952 barrels in 1906, and it continued upward for another three years. The Inland Oil Company was largely responsible for the turnaround. That corporation was organized in 1901 by officials of the United Oil Company, but the two enterprises were separate entities. Eight wells were completed in the years 1905 to 1907. The company's No. 13, known as "Big Liz," completed in July 1906, flowed at an initial rate of 200 barrels a day. In 1907, general manager W. R. Rathvon purchased a refinery at Lodi, Ohio, with a capacity of 500 barrels a day. It was erected adjacent to the Colorado and Southern Railway tracks, about midway between Boulder and Valmont. In February 1908, the plant made its first run at about half capacity, processing 300 barrels of oil a day. The Continental Oil Company distributed the products. For some time after that date, little if any oil was sent from the Boulder field to the United Oil Company's refinery at Florence.

Inland's refinery was, in fact, the field's second plant for processing crude oil. The first was built by R. H. Black, an experienced refiner from Pennsylvania. Black joined local entrepreneurs C. A. Dorn and Ed Wilson to establish the Boulder Oil and Refinery Company's refinery in 1902, at a cost of $10,000, on a site two miles northeast of Boulder. They planned to operate outside the Standard Oil monopoly, to buy crude oil at a dollar or more a barrel, substantially above the 84 cents a barrel paid by United Oil, and anticipated handsome profits from their venture.

Operations began in 1903 with a capacity of fifty barrels of crude oil a day. One month later, W. W. Degge, a prominent local realtor and president of the Wellington Oil Company, bought a controlling interest in the Boulder Oil and Refinery Company, and reorganized it as the Boulder Refinery Company, operator of a skimming plant that removed gasoline and kerosene from crude oil; the remainder was sold as boiler and furnace fuel. Degge's principal challenge was to overcome the near monopoly of the Continental Oil Company in the market for lamp oil and other petroleum products. He announced that local merchants who did not sell his firm's superior grade of kerosene had accepted rebates from the Standard Oil Trust. His assumption of the role of David in battling the Standard Oil Goliath may have been a good advertising ploy, but it was of short duration, for in 1906 Degge leased the refinery for a period of twenty-five years to the Inland Oil Company. That firm closed the small plant in order to eliminate competition with its own refinery.

In September 1909, the Inland company completed its No. 16 well, a large producer by the standards of the Boulder field. In addition to 450 barrels of oil a day, it and the company's other wells yielded daily about 700,000 cubic feet of "wet" natural gas. J. G. Johnson, the company's field manager, suggested that with a compressor, 3 gallons of gasoline could be wrung out of every 1,000 cubic feet of gas that escaped from the well. The gasoline could be sold to owners of automobiles, while the residual dry gas could be sold to the local gas company.

There was a growing market for gasoline in Boulder and adjacent communities. Largely for this reason, Inland's officials agreed with Johnson's recommendation to process wet gas from several gas wells located along a line that ran from a point about one mile east of Boulder in a northerly direction for about nine miles. When No. 27, completed in early September 1910, yielded an even larger volume of gas, the company installed machinery to extract gasoline from the flow. Inland's own wells produced about 1 million cubic feet of gas a day, and the Boulder-Greeley Oil Company another 600,000 cubic feet. The extraction plant commenced operation in 1911, handling about 500,000 cubic feet of gas daily. All of the firm's products were marketed by the Continental Oil Company, sometimes to the disadvantage of the United Oil Company's operations at Florence.

This prompted a reaction from a majority of United's stockholders, who had no financial interests in what was supposed to be an affiliated enterprise. In June 1908, Jacob Wallace, in behalf of fellow investors, initiated a suit in the Fremont County District Court against S. F. Rathvon, secretary-treasurer of the United Oil Company, and the officers of the Inland Oil Company for the recovery of $1 million in alleged misappropriated funds, and the appointment of receivers for both enterprises. The plaintiff charged that Rathvon and associates had improperly used United's funds and equipment to explore the Boulder field. After finding oil, they organized the Inland Oil Company and retained for themselves all of the stock. In addition, the suit alleged that the Inland Oil Company, acting through the Continental Oil Company, undersold United's products in the Colorado market.

In January 1910 the court ruled in the plaintiff's favor. Inland stock owned by the defendants was turned over to United, plus the sum of $14,671.25 per annum from April 1, 1908, as compensation for misused funds and equipment. United acquired three-quarters of Inland's stock, which gave Wallace, as general manager, control of the leading companies in Colorado's two commercial oil fields.

By that date, the Boulder field's output turned sharply downward, and continued to decline, almost without interruption, for several years. Without a large local supply of crude oil, the Inland company decided to close its refinery on July 1, 1911. Oil from the firm's wells was shipped again to Florence for refining, leaving the Boulder field with one operating refinery.

The Colorado–United States Refining Company operated a refinery at Loomis, seven miles northeast of Boulder in the northern section of the oil field, where most of the producing wells were located. Employing the Freeman process, designed to transform paraffin into oil, the plant, with a capacity of 600 barrels a day, completed test runs in November 1911. Initially, the declining output of crude oil hampered operations, leading to the sale of the plant to American Oil Processes Ltd. of London. In 1917, that company erected a tank farm and opened five service stations in Denver. A year later, the Apex Refining Company acquired all of the English firm's Colorado holdings. Apex planned to expand drilling activities in the Boulder field, and to process oil from its Big Hollow field near Laramie, Wyoming. The refinery continued operation until 1922.

The trend of output in the Boulder field was generally downward. There was a slight increase in 1919 and 1920, after which production dropped to 2,000 barrels in 1926. In all, 187 wells were drilled, only one-fifth of which were commercial producers. A high percentage of failures, and rapid depletion of producing wells, undermined the initial boom and discouraged efforts to revive the field in later years. A few wells, including the McKenzie No. 2, pumped small amounts of oil for many years.

The Florence field was the state's principal source of crude petroleum, as well as refined products, until the 1920s. The output in Fremont County enjoyed a brief revival in 1914, topping 200,000 barrels for the first time since 1909. Exploratory wells in Garfield, Mesa, and Rio Blanco Counties yielded small amounts of oil, but in the absence of economical transportation to large markets, there was no commercial production.

Prospectors, known as wildcatters, continued to search for new sources of crude petroleum. A wildcat well was one drilled where oil had not previously been known to exist. In 1915, exploration was conducted by locally organized companies in Adams, Crowley, El Paso, Garfield, Mesa, and Rio Blanco Counties. Two years later, similar efforts were reported in Alamosa, Arapahoe, Kit Carson, Larimer, Las Animas, Lincoln, Logan, Morgan, Otero, Pueblo, and Weld Counties. None of the exploratory wells opened new commercial fields.

Automobiles, because of their large appetite for gasoline, spurred the search for crude oil. In 1900 there were only 8,000 horseless carriages in the United States, and they were owned almost exclusively by the wealthiest elements of society. Ten years later there were 468,500. It was not until Henry Ford adopted an assembly line to mass-produce the Model T that the cost of motorcars came within the reach of a large segment of the American people. When that happened, the number of motorcars registered in the United States soared to more than 9 million in 1920, 20 million in 1925, and 26.5 million by 1930. The consumption of gasoline rose at a corresponding rate.

The number of registered passenger cars and trucks in Colorado increased from 13,135 in 1913, when the state assumed control of licensing, to 127,549 in 1920, and 308,509 in 1930. At the same time, the amount of gasoline purchased by Colorado operators of vehicles jumped from 5.8 million gallons in 1913, to 51.9 million gallons in 1920, and to 179.9 million gallons in 1930. Revenues derived by the state from gasoline sales rose from $274,401, for eight months in 1913, to $6.6 million in 1930. A tax of 1 cent per gallon to pay for highway construction was levied in 1919. It was doubled in 1923, with one-half devoted to highway construction, and the rest returned to the counties for roads. A 3-cent tax was imposed in 1927, with 70 percent assigned to the construction and maintenance of state highways and the remainder assigned to counties. In 1929, a new law assigned 70 percent to state highways, 27 percent to counties, and 3 percent to cities and towns for the construction and maintenance of streets.

Motorcars were instruments of sweeping change in American society. Their rapid acceptance after 1910 made inevitable the Good Roads Movement, and with the growth of highways, America experienced a rapid transformation. One aspect of change was the explosive growth of suburbs. In addition, an increasingly mobile way of life fostered new industries that catered to a nation of people on wheels. Business success was increasingly dependent on providing customers with spaces in which to park their cars. Tourism became a major growth industry. In time, there were drive-in restaurants, theaters, banks, and a variety of retail businesses. People reordered their habits around automobiles, which were both symbols of economic success and indispensable means of transportation.

Producers and refiners of petroleum products were obvious beneficiaries of the automobile age. The Continental Oil Company, as the region's principal distributor of petroleum products, moved in new directions, dictated by the changing character of the market. As an affiliate of Standard Oil's marketing empire, Continental was the region's leading distributor of kerosene. The voluntary dissolution of the Standard Oil Trust in 1892, due to adverse court decisions in Ohio and other states, had no effect on the western corporation's operations because John D. Rockefeller and his associates remained the principal stockholders in all Standard Oil enterprises. In 1899, Standard Oil of New Jersey reorganized as a holding company, which, unlike the trust, was legal per se because it was chartered in New Jersey, where corporations could legally own and control other corporations. Again, there was no outward change in Continental's operations.

Change was anticipated in 1911 when the U.S. Supreme Court ruled that the Standard Oil Company of New Jersey had illegally restrained trade, thereby violating the Sherman Anti-Trust Act. The Court, in 1913, ordered the monopoly's dissolution. At that time Continental lost all direct ties with the companies that had been part of the Rockefeller group. Thereafter, Continental functioned as an independent petroleum enterprise.

Up to that time, Continental had handled products refined in Colorado and Wyoming. Allegedly friendly agreements had diminished or eliminated competition, contributing to the firm's ability to pay dividends that ranged annually from 70 to 100 percent of its capital stock. The shares had been owned principally by John D. Rockefeller and his business associates. The dissolution order was implemented at a time when the demand for petroleum products was advancing at a rapid pace in the United States and abroad. By 1913, refineries produced not lamp oil, but gasoline for vehicles powered by internal-combustion engines.

A Continental Oil Company service station, one of many that opened in Colorado in the 1920s. The man in the photograph is using a hose to transfer gasoline from a gravity pump to his automobile. Gasoline was pumped into the graduated glass cylinder, from which it flowed by gravity into the car's tank. Courtesy, Denver Public Library, Western History Department.

As an independent enterprise, Continental Oil Company diversified operations to achieve balance between crude-oil production, refining, and marketing activities. In 1916, the purchase of United Oil Company gave Continental its first refinery, which was immediately expanded and modernized at a cost of $750,000. The subsequent acquisition of the Florence Oil and Refining Company boosted Continental's capacity to process crude oil from 3,000 to 4,000 barrels a day. Simultaneously, the company launched an aggressive exploration campaign in Colorado and elsewhere in the Rocky Mountain region to supply its refineries with crude oil.

Initially, exploration was conducted by subsidiaries. These were the United Oil Company; the Inland Oil Company and its successor, the Continental Oil Purchasing Company; and the Buck Creek Oil Company, which was active in Wyoming. Additions to the company's output of crude oil remained modest, however, until the 1920s.

Continental's marketing strategy shifted from tank wagons that dispensed oil products to householders to conveniently located outlets where motorists could purchase gasoline and other petroleum products for their motorcars. The company's first filling station was in Denver's warehouse district. Tanks of gasoline were positioned on a warehouse platform, allowing motorists to dispense fuel by gravity through a hose that ran to their cars. Continental's first service station, as opposed to a filling station, was erected in 1915. In Denver, three new boxy, terra-cotta structures began selling gasoline and automobile accessories to motorists. That design remained in use for twenty-five or thirty years, during which time pumps evolved from gravity-fed to electricity, and graduated gravity-flow cylinders were replaced with automatic gauges that displayed the amount of gasoline dispensed and its cost.

In spite of rising demand for petroleum products, the development of new oil fields in Colorado lagged well behind activities in adjoining states. The Salt Creek field made Wyoming the leader among petroleum-producing states in the intermountain West. Colorado Springs capitalists, whose wealth had been made largely in the great gold camp at nearby Cripple Creek, played key roles in the evolution of the industry. Verner Z. Reed acquired claims in the Wyoming field in 1910. He tried without success to interest officials of the United Oil Company in exploring the property for deep oil strata, which had the potential of being far more productive than the shallow wells active there for the previous twenty years. When that effort failed, Reed, with help from French and American investors, organized the Midwest Oil Company in 1911. That firm sank numerous successful wells, and laid a 6-inch pipeline from the field to a new refinery in Casper. Production rose to 12.4 million barrels in 1921, the year that Standard Oil (Indiana) bought an interest in Midwest's production and refining operations.

The success of Salt Creek stimulated exploration elsewhere in eastern Wyoming. Coloradans participated in the development of the Big Muddy field east of Casper. The discovery well was drilled by the Merritt Oil and Gas Company, a key investor in which was Albert Eugene Humphreys Sr., a Denver mining entrepreneur who had enjoyed considerable success exploring for oil in Oklahoma. Drillers employed by the Merritt firm tapped modest amounts of oil at a depth of about 1,000 feet. They continued downward in search of a larger flow. On November 1916, they reached a prolific oil sand at a depth of more than 3,000 feet. The well was a large producer. Humphreys's associates sold their interests to the Ohio Oil Company (a former Standard Oil affiliate), and in 1917, after holding out for a time, Humphreys transferred his interests to the Midwest Refining Company for $1 million in cash, plus stock. Following Humphreys's death in 1927, his two sons assumed direction of the family's mining and petroleum enterprises. Albert Eugene Jr., an astute businessman, and Ira Boyd, an inventor and mechanical genius, shared control, each contributing unique talents to a variety of undertakings. In the early 1930s, the brothers withdrew from the petroleum business, concentrating instead on the production of precious metals through the newly organized Humphreys Gold Corporation, leaving to others the task of finding oil pools in the intermountain West.

Much of the credit for big fields should be given to oil geologists, who, by the early twentieth century, were replacing the manipulators of "oil bobbers" and witching sticks as locators of oil pools. The Midwest Oil Company relied heavily on professionals for analyses of surface and subsurface structures as a means of eliminating some of the guesswork from searches. Cassius A. Fisher, who held baccalaureate and master's degrees in geology from the University of Nebraska, was instrumental in convincing Verner Z. Reed and his associates of the deep-horizon potential of Salt Creek. Carroll Wegeman succeeded Fisher as Midwest's chief geologist.

Professional geologist F. Julius Fohs was responsible for many of A. E. Humphreys's successes. Fohs had served as Kentucky state geologist from 1905 to 1912 before joining Humphreys and his associates in their search for oil in Oklahoma. It was at the urging of Fohs that Humphreys participated in the organization of the Merritt Oil and Gas Company to explore the Big Muddy field in Wyoming. Acting again on Fohs's advice, the Humphreys Oil Company drilled the first deep well at Mexia, Texas, in 1920, followed by similar successes in the nearby Currie, Powell, Bezette, and Richland fields. By 1925, when Humphreys withdrew from oil exploration, he had, with the assistance of Fohs, gained a reputation as one of the country's most successful wildcatters.

By the 1920s, geologists, who were called "rock hounds," "wrinkle chasers," and "pebble peeps," had been widely accepted as professionals who took much of the guesswork out of finding oil. Petroleum geology had advanced beyond the mapping of surface outcrops to new techniques for identifying subsurface features. Magnetic surveys, electrical impulses, radio waves, and shock transmissions greatly enhanced geologists' ability to identify underground petroleum reservoirs, with the result that large rewards often culminated searches for new fields.

Also in the 1920s, exploration in Colorado paid handsome dividends. Nine new fields, the first since Boulder, provided a surge of crude-oil production. In chronological order, they were the Wellington and Fort Collins, in Larimer County, completed, respectively, in 1923 and 1924; the Hamilton Dome, in Moffat County, and Tow Creek, in Routt County, in 1924; the Berthoud, in Larimer County, in 1925; North McCallum, in Jackson County, and Cañon City, in Fremont County, in 1926; Iles Dome, in Moffat County, in 1927; and South McCallum, in Jackson County, in 1928. The Wellington and Fort Collins, because of their close proximity, were eventually called the Fort Collins field. Cañon City was seen as an extension of the Florence field, and together they were known as the Fremont County field. The state's output of crude oil had declined early in the 1920s, revived in 1924, and exceeded 1 million barrels for the first time in 1925. From 1926 through 1929, production surpassed 2 million barrels a year.

Development occurred at a fast pace for many reasons, such as important advances in techniques of geological exploration and the fast-growing market for motor fuel. Contributing to the surge in production was a federal law that provided tax incentives to encourage exploration and development. The law authorized tax write-offs for the principal costs of putting down a well, and a large part of its subsequent income. Congress had adopted this policy in 1918, convinced that steps had to be taken to encourage the discovery and production of oil, as well as other fuel resources, to support the war effort against Kaiser Wilhelm and Germany's allies in World War I. This so-called depletion allowance, with only slight modification, remained in effect until 1986.

The Wellington–Fort Collins field completely eclipsed the production of Florence and Boulder. In fact, initial exploration in Larimer County, where the city of Fort Collins was the county seat, had occurred simultaneously with the development of the Boulder field at the turn of the century. The well-drilling technology of that day could not reach sufficient depths to find commercially important pools of oil. In 1916, S. F. Rathvon organized local capitalists as the Poudre Oil and Gas Company to make a deep test on what was supposed to be a promising geological formation known as the Wellington dome. The site was selected by Dr. R. D. George, the state geologist and an early champion of drilling identifiable geological structures. At a cost of $25,000, the company pushed the discovery well to a depth of 4,000 feet without finding oil. The duster was a severe blow to George's reputation until it was revealed that the site he had recommended was six miles north of the test well.

Convinced that structures were keys to successful exploration, geologists employed by the Roxana Oil Company, a Tulsa, Oklahoma, subsidiary of Shell Oil, reexamined the Fort Collins area in 1917. They identified three closed domes, on which the firm obtained leases for tests in 1919. Drilling was terminated at 1,300 feet, against the advice of company geologists, who recommended going to greater depth. The result was another dry hole. Roxana's effort had attracted the attention of other firms, so for some time the duster discouraged additional tests in Larimer County.

In 1922, Union Oil Company of California, following a careful reexamination of the geological formations, acquired leases for tests. Working on the Wellington dome, drillers penetrated oil-bearing sand at a depth of 4,285 feet on November 11, 1923. The well produced 82 million cubic feet of natural gas in twenty-four hours. It blew for forty days before crews could shut it down. During that time, it spewed a considerable quantity of high-grade crude oil.

The huge "gasser" launched a boom in the Fort Collins area. The Union Oil Company spudded four more wells in the spring of 1924. Two were on the Wellington dome, one on the Fort Collins dome ten miles to the south, and another on the Douglas Lake dome, six miles southwest of the initial test site. The completion on the Wellington structure produced gas, with some oil. The hole on the Fort Collins structure yielded a large quantity of crude petroleum. The volume remained unknown for a time because it was plugged until storage tanks could be erected. Production peaked at 1.22 million barrels in 1926. The Douglas Lake test was a dry hole.

Union's successes ignited an "oil rush" in northeastern Colorado. Leases were secured for an estimated 400,000 acres at an average cost of about 20 cents an acre. Many ventures were floated, but prospects were not bright because the California firm had secured practically all acreage on the structures. Drilling on adjacent lands entailed high risks of failure. Nonetheless, many oil enterprises sold stock with par values of 10 cents or less per share. The low values encouraged investors to buy into exploration companies, which greatly enhanced the profits of promoters, most of whom had difficulty raising the $100,000 or so required to sink a well to a depth of

The Continental Oil Company's headquarters building, at Eighteenth Street and Glenarm Place, Denver, opened in 1927. When the Marland Oil Company absorbed the Denver firm in 1929, it kept Continental's name, but moved the headquarters for combined operations to Ponca City, Oklahoma. The building served as Continental's regional headquarters until 1977, when the offices were moved to the new Anaconda Tower. The building was leveled and the site used for a parking structure to serve the Anaconda Tower and the Fairmount Hotel. Courtesy, Denver Public Library, Western History Department.

4,000 feet or more. Not surprisingly, therefore, all of the area's successful wells were completed by the Union Oil Company.

While attention focused on the Fort Collins area, the Texas Production Company, the Texas Company's (today's Texaco) subsidiary, initiated a test on the Hamilton dome (later called the Moffat field) near Craig. Work on the well commenced on June 23, 1923. Drillers encountered some oil in shale at a depth of about 2,000 feet. They eventually tapped a productive oil sand at 3,280 feet on March 3, 1924, bringing in a well that flowed at 4,800 barrels a day. Production was piped twenty-two miles to Craig, where a tank farm provided temporary storage. The oil was shipped via the Denver and Salt Lake Railroad (Moffat Road) to Denver, where it was rerouted to refineries at Florence and Parco, Wyoming.

The Moffat field was the first of several located in Colorado's northwestern quadrant. Texas Production sank the discovery well in the Tow Creek field, west of Steamboat Springs, in 1924. In quick succession, Midwest Refining Company opened the Garmesa gas field straddling the Garfield and Mesa county line, and the Iles Dome, a gas field west of the Moffat field, in 1924. The Gypsy Oil Company developed the oil potential of Garmesa in 1926. From 1926 to 1928, Continental Oil found petroleum and a large volume of carbon dioxide in the two McCallum fields. The Midwest Refining Company brought in a 1,900-barrrel well in deep sand on the Iles dome in April 1927.

By the close of the decade, Colorado's oil production was largely from Western Slope counties. The Moffat, Iles, and Tow Creek fields were large producers and, with the output of shallow wells at Rangely, accounted for 62 percent of the state's production in 1929.

As the production of crude oil increased, Colorado's refinery capacity expanded to keep pace with the growing market for petroleum products. The largest producer of gasoline, kerosene, fuel oil, and lubricants was the Continental Oil Company. It was twice absorbed by other corporations, first by the Mutual Oil Company in 1924, and again by the Marland Oil Company in 1929. In each instance, the takeover corporation chose to adopt the Continental name because of its long-standing prominence in the region west of the Mississippi River.

The Continental Oil Company acquired from each merger enlarged capacity to produce, refine, and market petroleum products. Mutual owned producing wells in Wyoming and in the Mid-Continent field of Kansas, Oklahoma, and Texas. Refineries at Glenrock, Wyoming, and Chanute,

Kansas, turned out products that were distributed by almost 400 filling and bulk stations in Kansas, Nebraska, Missouri, South Dakota, and Colorado.

After Mutual assumed control of Continental, the company conducted a sizable exploratory campaign in Colorado. From 1925 to 1927 it put down wells in Custer County south of the Florence field. One was dry, another encountered only a show of oil, and a third found oil at a depth of 2,530 feet, but it had an output of only five barrels a day. From 1924 to 1929, new wells near Cañon City contributed to the output of the Fremont County field. A 1924 wildcat well in Yuma County, designed to test deep sands for oil, found natural gas, but not in commercial quantities. That year, a test hole in Pueblo County was abandoned at 1,425 feet without finding oil.

Continental's most unusual exploration was in Jackson County. In what was called the North McCallum field, a test well struck gas and oil at a depth of 5,146 feet in 1926. Thirty million cubic feet of carbon dioxide a day carried with it 450 barrels of oil. The two came to the surface at a temperature of minus 110 degrees Fahrenheit. Efforts to separate the two into commercially marketable products were unsuccessful at that time. For that reason, when a 1928 test well in the South McCallum yielded carbon dioxide and oil, it was shut down. A second North McCallum well in 1928 produced 50 million cubic feet of carbon dioxide and thirty barrels of oil a day, which had no commercial value.

Outside of Colorado, Continental developed large production and refining capacities in the Southwest. From 1922 to 1925 the company, along with Midwest Refining and Santa Fe Mutual, opened the Hogback, Rattlesnake, and Table Mesa fields, the first commercial production in New Mexico's portion of the San Juan Basin. A refinery on the east side of Farmington, a joint venture with Santa Fe Mutual, processed the light oil. Continental was one of the first companies to develop the Artesia field of Eddie County in southeastern New Mexico. A refinery at Artesia handled the output.

The Marland Oil Company owned numerous wells, plus two refineries, in Oklahoma. Its plant near Ponca City, the location of the corporation's headquarters, was the largest in the Mid-Continent field. Marland had played, prior to 1929, only a small role in Colorado. A 1924 well in the Meeker field, of Rio Blanco County, tapped an immense flow of water at 3,400 feet and was abandoned. In 1925, and again in 1927, the company completed successful natural-gas wells in the Thornburg field of Moffat

County. In the latter year, a well on the Iles dome, in the same county, reached a depth of 3,400 feet without finding oil.

As a result of the second merger, the Continental Oil Company acquired a new home. Its headquarters was relocated from Denver to Marland Oil's main offices in Ponca City. The Denver office was relegated to regional administration. The reorganized enterprise was a major producer, refiner, and distributor of oil in the trans-Mississippi West, particularly the high plains from Texas to South Dakota. Six refineries featured state-of-the-art technology, including distillation towers, continuous chemical treatment, quality-control laboratories, and modern cracking stills. Plants at Wichita Falls, Texas; Sapulpa, Oklahoma; Glenrock, Wyoming; Artesia and Farmington, New Mexico; and Florence, Colorado, were, with one exception, near large sources of crude oil and major urban centers, important markets for petroleum products.

The Colorado plant was the exception. It had a capacity of 1,500 barrels a day. Lighter fractions, including gasoline, kerosene, wax, and lubricants, were removed by heating crude oil in stills. The residue was transferred to Burton cracking stills, which had a capacity of 1,500 barrels a day. They reduced heavy oil to lighter fractions, including gasoline. The Florence field, however, was incapable of meeting all of the refinery's needs for crude oil, which it secured from western Colorado and Wyoming. Florence was more than a hundred miles from Denver, the region's largest market. It came as no surprise when the corporation announced plans for a new state-of-the-art refinery in the Denver metropolitan area in 1929. To gain an ensured supply of crude oil for the new plant, Continental Oil purchased all of the Union Oil Company's oil holdings in the Wellington–Fort Collins field.

The Florence plant was the largest of several refineries active in the state in the 1920s. The Apex Refining Company operated a 200-barrel plant at Loomis from 1918 to 1921. It was a skimming operation that removed from crude oil only highly marketable components—gasoline and kerosene—and sold the rest as boiler or furnace fuel. Another skimming plant was operated at Rangely by the Raven Oil and Refining Company. The 100-barrel unit processed the product of the area's numerous shallow wells. Opened in 1920, it remained in operation throughout the decade, supplying the local market with gasoline and illuminating oil. In 1925, the Anaconda Oil and Refining Company planned to open a skimming plant with a capacity of 250 barrels a day at Genoa, in Lincoln County.

This was in anticipation of production that did not materialize. In 1925, Texaco erected a state-of-the-art refinery at Craig, complete with Holmes-Manley cracking stills, to treat the output of its wells in Moffat and Routt Counties. The plant had a capacity of 1,500 barrels a day, and shipped petroleum products to market in the Moffat Road's rail tank cars.

The decade closed with the Colorado oil industry in a healthy condition. However, the Great Depression was to have a negative impact on the production, processing, and distribution of petroleum and petroleum products. Emergency measures in 1930s sustained exploration and development until military and industrial mobilization during World War II set the stage for another round of expansion in the postwar years.

4

Lighting and Heating with Gas 1870–1930

By the late nineteenth century, gas, either natural or manufactured, was an alternative to coal and oil for heating and illuminating homes and businesses in many states. Natural gas, composed largely of methane with small amounts of hydrogen and nitrogen, was, if mixed with air, highly explosive. That was also true of manufactured gas, made by the destructive distillation of bituminous coal. It was almost pure methane.

Although natural and manufactured gas were interchangeable fuels, they differed in two important respects. Natural gas was relatively inexpensive, and it had high heat values, measured in British thermal units (Btu). By contrast, manufactured gas was expensive to make, and its heat values were low, often 50 percent, or less, of natural gas. Consequently, natural gas, if available, was preferred over manufactured gas. In Colorado, as in much of the nation prior to the second quarter of the twentieth century, natural gas was either nonexistent or in short supply. Therefore gas, if available for heating and illumination, was of the man-made variety.

Gas distilled from coal in airless retorts was first used in Europe to create a convenient fuel for illuminating streets during hours of darkness. This process was first applied in the United States in 1816 to produce gas for Baltimore's streetlamps. Later, manufactured gas was distributed through underground mains to residential and business customers in Baltimore.

Similar events subsequently occurred in many towns and cities throughout the United States.

In 1900, Colorado was credited with a very small, almost minuscule, production of natural gas. Wells in the oil field at Florence contributed only 0.006 percent of the nation's total output. In other words, there was at that time little production worthy of notice. That would change in time, but until then, people who were willing to pay for the convenience of gas for lighting or heating had to purchase the manufactured variety and, at least in Colorado, they could do so only if they resided in the state's principal urban centers.

In 1902, the first year for which there are reliable statistics for the production and consumption of manufactured gas in the United States, there were 533 plants that distilled gas from coal. Their total output was 30.8 billion cubic feet, four-fifths of which was consumed for illumination, at an average cost to consumers of $1.01 per 1,000 cubic feet. Most of the gas was burned in streetlamps. Only 20 percent was sold to households and businesses. Utilities, by selling valuable by-products of the distillation process—namely, gas ammonia liquors, tar, and gashouse coke—added to income derived from the sale of gas.

In the Rocky Mountain region there were twelve coal-gas plants in 1902. Five were in Colorado; five more in a three-state region comprising Montana, Nevada, and New Mexico; and one each in Utah and Wyoming. Colorado's retorts, located in Denver, Leadville, Gunnison, Pueblo, and Colorado Springs, distributed to customers 393.1 million cubic feet of gas for lighting and heating. Coal gas, although widely used, had a limited impact largely because its manufacture and distribution, given the high cost of facilities, were restricted to well-to-do residents of urban centers at a time when most Americans had modest means and often lived in rural villages and hamlets, or on farms.

The largest and oldest of Colorado's gas plants was located in Denver. The community was little more than a year old when the Arapahoe County Gas Light Company sought a franchise to make and distribute illuminating gas. The venture, and five more over a period of nine years, proved to be premature. It was not until 1869 that James Archer, with local investors, organized the Denver Gas Company. A native of Belfast, Ireland, Archer had acquired a sizable fortune in the iron trade at St. Louis before relocating to Denver. With the completion of the Denver Pacific Railroad to Cheyenne and the construction of the Kansas Pacific to Denver from the

East, he decided to invest in the future of a city that seemed to be poised for dramatic growth. He prevailed upon a number of men to join him in forming a gasworks to generate fuel for lighting streets, as well as for illuminating and heating homes and businesses.

The investors were distinguished community leaders. They included William N. Byers, editor of the influential *Rocky Mountain News*; Louis Bartels, a grocer; Walter Cheesman, a druggist; Louis N. Tappan, a hardware merchant; David Halliday Moffat Jr., a banker; D. Tom Smith, agent for a freight line; and John Evans, railroad builder and former governor of Colorado Territory. With two exceptions, these men were important shapers of Denver's growth. Tappan soon returned to his native Massachusetts, and Smith left Denver, and the country, to avoid prosecution for shooting a prospector during a barroom brawl.

The Denver Gas Company, with Archer as president and Byers as secretary, was organized on November 14, 1869. Under terms of a twenty-five-year franchise from the City of Denver, the corporation generated and distributed coal gas for streetlighting commencing in 1871. The municipality furnished the iron posts and lamps, and paid the utility $55 per unit a year for fuel and maintenance. Lamplighters patrolled the streets at dusk with ladders and matches to ignite the burners, and returned at dawn to extinguish the flames. The company's retorts were located at Wewatta and Eighteenth Streets, later part of the rail yards in lower downtown Denver. In 1872, the Denver Gas Company produced about 13 million cubic feet of gas that circulated through nine and a half miles of underground mains within the boundaries of the city. By 1873, 10,000 gas burners, mounted on iron posts, illuminated the streets at night.

Customers paid a stiff price to illuminate homes and commercial properties. Householders paid $5 per thousand cubic feet of gas until 1877, when the price fell to $3 per thousand cubic feet. At a time when unskilled workingmen earned no more than a dollar a day, the price of manufactured gas was beyond the means of most of the city's residents.

Denver's streets were illuminated with gas only a short time. In 1883, the gas lamps were supplemented with arc lights positioned at the top of 150 towers at strategic points throughout the community. The superiority of electric lights, as opposed to gas burners, was quickly apparent. Bowing to public opinion, the city abandoned gas lamps for its streets.

The Colorado Electric Company, organized in 1881, won the contract from the City of Denver, offering a less expensive and more efficient method

Denver Gas Works, located at Eighteenth and Wewatta Streets, 1915. The plant, erected in 1870, distilled coal to produce what was called "manufactured" or coal gas, which was used to illuminate the city's streets, homes, and businesses. The gas retorts were shut down in 1928 when the Public Service Company of Colorado secured natural gas by pipeline from Texas. Courtesy, Denver Public Library, Western History Department.

of illuminating streets. That enterprise was formed by William G. Fisher, pioneer Denver merchant; Walter A. Cheesman; Frederick O. Vaille, organizer of the local telephone exchange; Nathaniel M. Tabor, son of pioneer H.A.W. Tabor; and others. Using Brush arc lamps, the firm offered to light the city's streets at a cost of $6,500 a year, well below the gas company's bid of more than $11,000 for the same service. Electric lights were installed in 1885 when the Denver Gas Company's contract with the city expired.

As the community grew, the city council awarded franchises to other enterprises. The United Improvement Company and the People's Gas and Light Company, in 1883 and 1887, respectively, introduced substantial reductions in rates that the Denver Gas Company had to match. Competition ended in 1891 when the three firms merged to form the Denver Consolidated Gas Company.

Gas lamps lost favor with most Denver residents after the electric utility replaced direct with alternating current in the 1890s. Man-made gas gained popularity, however, as fuel for cookstoves. The retort house was expanded from time to time until natural gas from Texas entered the Denver market in the late 1920s.

The Leadville Illuminating Gas Company was formed in April 1879, at the time of the silver-mining boom. The principal organizers were H.A.W. Tabor, William H. Bush, and C. D. Hall. Under the supervision of general manager D. H. Ireland, the company constructed retorts, a gas-holding tank, and underground mains to commercial establishments and private homes. The Tabor Opera House opened that year with gaslights. One hundred lamps were installed on the streets. The fuel consumed in the retorts was bituminous coal from Trinidad.

Gunnison, a railroad town and commercial center of 5,000 population in the valley of the Gunnison River, had an extensive system of gaslighting. B. W. Lewis, a St. Louis entrepreneur, organized the Gunnison Gas and Water Works Company in 1881. David J. McCanne resigned as manager of the Lewis Iron Works in Grand Tower, Illinois, to relocate to Gunnison, where he took charge of building and operating retorts and a waterworks. Gas was used for lighting streets and distributed to homes and businesses. The company was also a wholesale and retail dealer in machinery, steam fittings, plumbing and gas fixtures, superior domestic-grade coke, tar, pitch, and roofing paints. The last four items were products of the gas retort.

Pueblo's emergence as an urban-industrial center in 1881 prompted rival firms to seek gas franchises from the city. The winner of a short but spirited contest was the Union Gas Company, organized by Mahlon H. and John D. Thatcher, prominent bankers, with the participation of many local investors. The enterprise built retorts and installed forty-six lamps on posts at street intersections. The burners were ignited for the first time on November 26, 1881. Business flourished as residents contracted for gas connections for homes and offices. In 1901, the ownership of the firm passed to Chicago capitalists.

The Colorado Springs Gas and Coke Company secured from the city a twenty-five-year franchise in July 1879 to generate and distribute manufactured gas within the municipality's boundaries. In January 1880, streetlights were ignited, and gas was supplied to thirty stores and seventy-five other structures. The Lowe Gas and Electric Company bought the franchise, plant, and distribution system in March 1881. That enterprise sold out to the Colorado Springs Gas and Electric Company in 1894. Five years later, it was purchased by Bonbright and Company, an operator of utilities in cities throughout the United States.

Advocates of electricity predicted that it would quickly displace gas in Colorado. However, both utilities prospered, and in time merged to form single enterprises. Electricity was adopted for lighting, while gas was used for space and water heating, and for cooking.

In addition, retorts produced, as by-products of the gas-making process, ammonia liquor, tar, and coke. Ammonia liquor was treated with sulfuric acid to make ammonia sulfate, a popular and inexpensive fertilizer. Tar was widely used to make roofing materials. Gashouse coke, so named to distinguish it from the coke used in blast furnaces, was an ideal household fuel for urban dwellers because, unlike some grades of coal, its combustion emitted little smoke and soot.

Initially, Colorado operators of retorts purchased gas-making coal from several mines in Wyoming and Colorado. In time, experience revealed that the metallurgical coal of the Trinidad district produced gas that was superior in quality and quantity to that of other fuels. It was readily available at nominal cost following the completion of the Denver and Rio Grande Railway to El Moro and to Trinidad. After the same narrow-gauge line built a branch from Gunnison to Crested Butte, the latter community's high-grade metallurgical coal became the fuel of preference for gas companies throughout Colorado and nearby states.

In the first decade of the twentieth century, additional Colorado towns and cities turned to manufactured gas for illuminating streets at night, while at the same time providing residents, if they paid for the service, fuel for lighting and heating. Franchises, in fact monopolies, were given to private enterprises to manufacture and distribute gas within the confines of municipalities. Ordinances authorized the construction of mains under streets and alleys to carry gas to homes and businesses. Itinerant promoters offered to assist local capitalists to organize, build, and operate gas utilities. Boulder was an example of this method of developing municipal utilities.

In February 1902, three residents of Titusville, Pennsylvania, proposed to build a community gas system to serve the city of Boulder. The promoters—Charles C. Hurd, Henry H. Hurd, and Peter English—recruited local investors with whom they formed a corporation—the Boulder Gas, Light and Fuel Company—which asked the city council for a franchise to make gas from coal and oil and to sell it to residents for lighting, cooking, and space heating. The council granted a twenty-five-year franchise, contingent on the start of construction of a retort and underground mains within three months. That deadline passed without the requisite action because the Hurds did not, as promised, provide the $25,000 needed to launch the company.

A new corporation was quickly created. Peter English and his son Frank joined local businessmen in forming the Boulder Gas Company. The president and principal investor was Andrew J. Macky, head of the First National Bank of Boulder. He was joined by two members of the city council, Eugene A. Austin and J. B. Teagarden. The city council issued a new franchise authorizing the enterprise to erect a gashouse on Thirteenth Street, adjacent to the Colorado and Southern Railway tracks in downtown Boulder, and to build underground mains to deliver gas to streetlamps, businesses, and residences.

The plant was designed by Peter English, who claimed to own patents on processes for generating gas from crude petroleum and coal. English came to Boulder from Benton Harbor, Michigan, where he had recently erected a gas plant in cooperation with local investors. He had built similar works throughout the Midwest, and his reputation for fulfilling all terms of contracts with local interests was hailed as assurance that Boulder's gas-making project would be successful.

Gas-making commenced in December 1902, but the company did not enjoy quick success. For a time it operated jointly with the Boulder Electric Light and Power Company, a union that ended when the gas company's directors ordered an intensive campaign to sell gas for illumination, an action that seemed to target the existing market for electric lights. The directors believed the gas company had advantages over alternative fuels because it charged only $1 per 1,000 cubic feet of gas used for lighting, cooking, and heating. Gas was cheaper than electricity, and both were less expensive and more convenient than coal.

The gas company, even before it had a chance to prove its advantages over other forms of energy, faced the threat of competition from natural gas. The National Consolidated Oil Company, composed of Buffalo, New York, investors, asked the Boulder City Council for a franchise to distribute natural gas to households and businesses. The same Buffalo men were active in the beet-sugar company at Sugar City in the Arkansas River Valley. In Boulder, they proposed to sell natural gas at rates as low as 30 cents, and not more than 75 cents, per 1,000 cubic feet. This was less than the Boulder Gas Company's cost of manufacturing and distributing gas. The city council approved a franchise authorizing the company to distribute natural gas when it became capable of delivering 150,000 cubic feet of gas a day. Its one source, popularly called the Buffalo well, began with an output of 50,000 cubic feet per day. However, the volume increased rapidly to, and beyond, 150,000 cubic feet per day.

Located on George B. Williamson's ranch one mile east of Boulder, beyond the bridge that carried Arapahoe Road (later Colorado Highway 7) over Boulder Creek, the well was far to the south, but in the same alignment as the field's principal oil and gas producers. National Consolidated held a twenty-year lease on the land where the well was located. The lessees tried without success to complete additional gas wells in the vicinity of their one producing unit.

The Boulder Gas Company, faced with potentially ruinous competition, sought an agreement with National Consolidated. The well's owners, hoping to avoid the cost of constructing an underground system of mains to distribute gas in Boulder, entered into a twenty-year contract to sell natural gas exclusively to the current holder of the community gas franchise. The Boulder Gas Company, for its part, was committed to using natural, rather than manufactured, gas as long as the supply was adequate to meet the community's needs.

The gas was transmitted to Boulder by pipeline and, at least for a time, mixed with manufactured gas for distribution to customers within the community. The rising output led, in September 1903, to the company's decision to halt the manufacture of gas. Subsequently, the plant remained on standby status in the event that demand exceeded the supply of natural gas.

The dry natural gas was odorized, as required by law, but otherwise sold without processing. The Boulder Gas Company purchased only about one-half of the approximately 3 million cubic feet of gas produced monthly by the Buffalo well. This was because the 2-inch line that linked the well to distribution mains in the city could not carry all of the output. A new 4-inch line in 1905 increased the flow from the well to the city, but because demand did not advance as fast as supply, a large amount of gas was allowed to escape into the atmosphere.

The Boulder Gas Company, reorganized as the Federal Gas Company in March 1904, sought by various means to increase its sales in the community. It extended mains to University Hill, Mapleton Hill, and other sections of the city in an effort to attract new customers. In addition, it offered what proved to be popular cooking classes to acquaint housewives with the advantages of gas over electric, coal, and wood ranges. Representatives were dispatched to homes to demonstrate how to get the best results from gas appliances. An advertising campaign informed Boulderites that gas burners installed in place of coal grates in furnaces were cheaper and cleaner to operate, with the further advantage of big savings in labor and maintenance.

The campaign was successful. Customers increased in number from 800 in 1907 to 1,000 in 1908. In the latter year, the city was served by eighteen miles of underground mains that annually delivered about 30 million cubic feet of gas to customers. This was an increase of 8 million cubic feet over the previous year.

The supply of natural gas from the Buffalo well gradually increased over the years. The output rose from about 30 million cubic feet a year in 1904 to about 35 million cubic feet in 1909. However, attempts to sink gas wells in the same area east of Boulder were uniformly unsuccessful. Farther north, several wells, particularly those of the Inland Oil Company, produced large volumes of gas, but no effort was made to pipe it into the city. The oil-well gas was processed to extract natural gasoline, and the residue was either burned under boilers to generate steam for operating drilling rigs and pumps or vented into the atmosphere.

In 1914, the Buffalo well ceased production when it was flooded by water from nearby Boulder Creek. Frantic efforts to reestablish contact with the underground reservoir or to secure supplies elsewhere in the Boulder field were to no avail. In order to retain its franchise, the Federal Gas Company built a new generator designed to produce daily 350,000 cubic feet of gas.

By then, Peter English no longer resided in Boulder. After serving as vice-president of the Boulder Gas Company for about a year, he sold his stock in the enterprise and turned once again to building gasworks in communities throughout the Midwest. In 1904, with son Walter, he purchased a gas plant in Deadwood, South Dakota. Later that year, he built a gashouse in a suburb of Chicago, and also opened negotiations for the purchase of the gas and electric plants in Greeley, Colorado. Another son,

Frank, who was in charge of a plant in Watertown, South Dakota, asserted that his father had dozens of contracts to build new gasworks.

While the Federal Gas Company returned to manufactured gas, the National Oil and Gas Company, lessees of the well on the Williamson ranch, continued efforts to revive the Buffalo well, or to sink others of equal worth in the vicinity. This activity yielded no tangible results from 1914 through 1918. In 1919, the Northern Gas and Drilling Company sank a new well only twenty feet from the original hole. At a depth of 1,500 feet drillers encountered gas, but continued to about 1,630 feet, the depth at which gas flowed from the Buffalo well. In April, the drilling crew encountered a flow of about 50,000 cubic feet a day, approximately the initial yield of the old well. Heartened by this development, the company anticipated the flow would expand. They were disappointed. In 1930, when natural gas again flowed to customers in Boulder, it was transported by pipeline from the Texas Panhandle.

By about 1910, manufactured gas had reached the limits of its growth as a fuel for lamps. Householders in Colorado, as elsewhere in the United States, discovered that electricity was a more efficient, if somewhat more expensive, source of lighting, and that its mechanical and industrial applications were almost without limit. By that date, most gas companies, as suppliers of fuel for cooking and heating appliances, had been absorbed into corporate combinations that served as local and regional light, power, and heating utilities.

Gas lights retained some popularity until well into the twentieth century, largely because of the extraordinary efficiency of the Welsbach mantle. The invention of an Austrian chemist, Carl Auger, Baron of Welsbach, the mantle was made of gauze impregnated with oxides of thorium and cerium. When heated in a gas flame, the mantle became incandescent, emitting a brilliant light. It increased the illumination value of gas from two or three candlepower, as a naked flame, to about twenty candlepower. Welsbach established a factory in London in 1877 to manufacture the mantle for household use. He developed a mantle for streetlights in 1895. Long after electricity had captured the indoor- and streetlighting market, the Welsbach mantle, or variations on it, continued in use for outdoor and camping lanterns.

By 1920, all states, plus the District of Columbia, had facilities for manufacturing and distributing gas, which was consumed almost exclusively for cooking and heating. That year, 8.8 million customers in approximately 4,000 towns and cities of the United States purchased 311

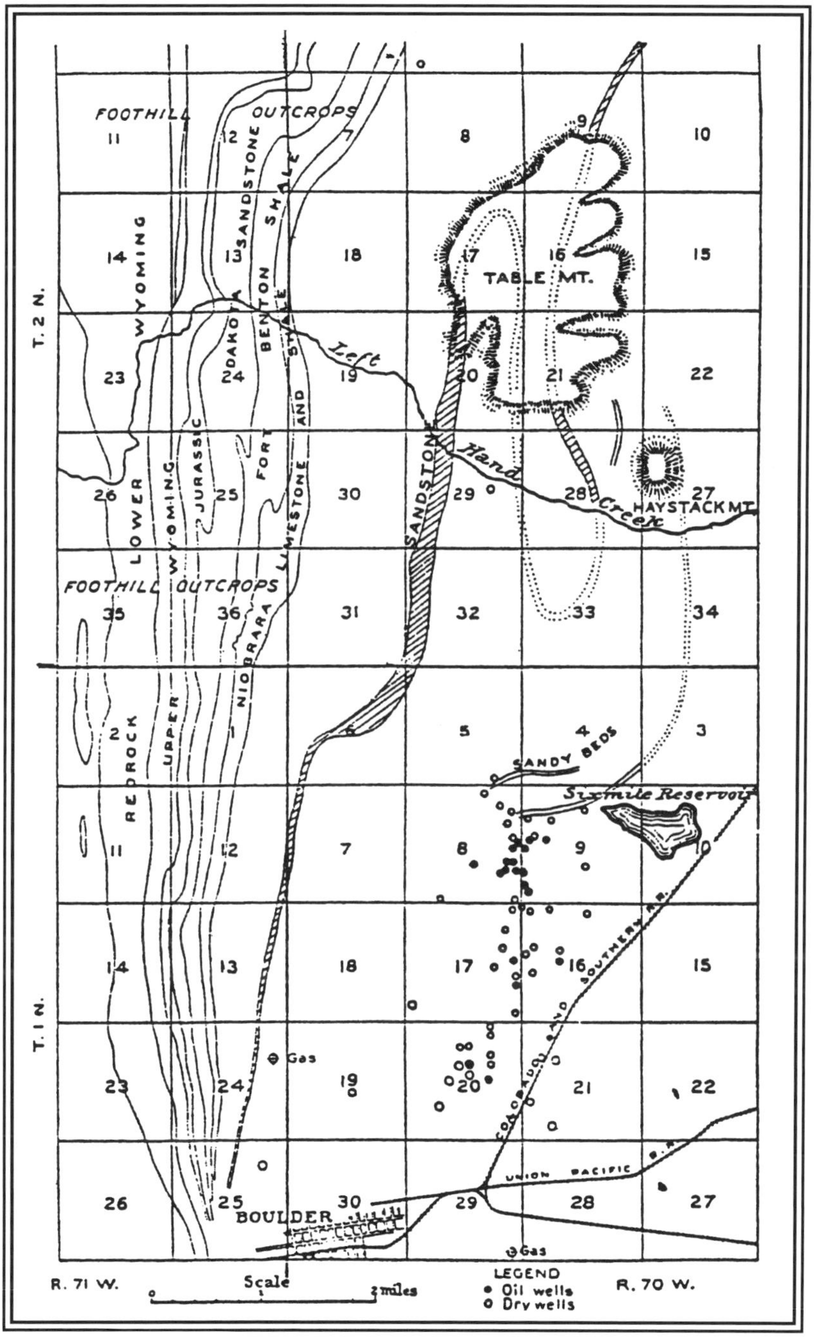

Map 3. Boulder Oil and Gas Field, 1903. From USGS, *Bulletin 225* (1904).

billion cubic feet of gas. Although gaslighting had been largely superseded by electric lights, almost 46 million people enjoyed the convenience of gas for cooking and for heating homes and businesses. The cost of manufactured gas precluded its consumption for space heating in most working-class households.

In 1920, Colorado had fifteen gashouses that produced a total of 10.8 million cubic feet of gas. Plants were located principally along the Eastern Slope from Trinidad in the south to Fort Collins in the north. Boulder, Cañon City, Colorado Springs, Denver, Greeley, and Pueblo were among the cities whose residents enjoyed the convenience of gas. Denver, the state's principal population center, had the largest facilities for making and distributing gas within the limits of the city's boundaries. On the Western Slope, Leadville, Durango, Gunnison, and Grand Junction had gasworks.

Colorado's gas plants were of four basic designs—seven distilled coal in retorts, six produced water gas, one used petroleum, and another made industrial coke for reducing metallic ores. Ovens for making smelter and blast-furnace fuel were outwardly similar to retorts for making coal gas. Commencing in 1918, the Colorado Fuel and Iron Company activated the state's first and only set of by-product ovens at its Pueblo steelworks where, in addition to coke, the company reclaimed ammonia liquor, benzol, and tar from oven gases. Coke-oven gas was similar to coal gas, except that it was a secondary product of ovens designed primarily to yield high-quality coke for industrial consumption.

Water gas was manufactured by passing steam over a bed of white-hot coals. This was an intermittent process. Initially the temperature was raised by forcing a blast of air through the burning fuel, after which steam was applied. Water gas had lower heat values, measured in Btu, than coal gas, but the cost of building and operating a plant was significantly lower than for a retort. Furthermore, production could be easily adjusted to reflect changes in demand, which in most municipalities fluctuated daily, weekly, and seasonally.

Oil gas was made by substituting petroleum for coal. Because of the abundance of coal and the relatively higher cost of oil, this method was employed in only one Colorado community, which was not specified in the published data of that time. As with water gas, oil gas could be adapted readily to fluctuating levels of demand.

Except in a few eastern and midwestern states, public and private utilities that manufactured and sold gas experienced little competition from natural gas until the decade of the 1920s, even though large untapped reservoirs abounded in many parts of the country, particularly in Colorado, Kansas, Oklahoma, Texas, and California. In many parts of the Rocky Mountain region, including Colorado, the absence of major population centers meant that there was relatively little local demand for the fuel.

Natural gas had been known to exist in eastern states for many years before it was piped from the ground to illuminate thoroughfares and households. Its earliest recorded use occurred in Fredonia, Chautauqua County, New York. In 1821, shallow wells put down on a seepage collected gas, which was distributed through wooden mains to light the streets. Each lamp produced a pale flame, a low-intensity light equivalent to about two candlepower. It was not until 1865 that the Fredonia Gas, Light and Water Works Company distributed natural gas to homes. Other communities followed Fredonia's example, tapping into underground reservoirs of gas for lighting and heating.

Wherever gas was abundant, manufacturers consumed it rather than coal. Gas, with high heat values and low impurities, was an ideal fuel for the metallurgical, glass, pottery, and cement industries. These flourished in western Pennsylvania and Ohio until gas fields declined in output, which prompted either a change to alternative fuels or relocation to new gas fields. Because of the critical importance of fuel without impurities, glassmakers frequently moved in search of new supplies of natural gas.

Viewed as an inexhaustible resource, natural gas was shamefully wasted for many years. Householders were charged fixed monthly rates, without regard for the amount consumed. Manufacturers purchased gas on contracts based on the value of goods produced, not the amount of energy consumed. Dwindling supplies usually forced producers and distributors to introduce meters in an effort to encourage conservation. Waste was effectively discouraged when gas prices matched the cost of alternative fuels.

In the nineteenth century, Colorado's principal source of natural gas was the Florence oil field. All wells produced gas, but the amount was usually small, often no more than the volume of oil. Ten wells produced only gas, which the operators converted, at first, to their own use. It was an ideal fuel under boilers to produce steam for operating pumps to bring oil to the surface. In 1887, gas not needed in the field was transmitted via pipeline to Florence to illuminate the United Oil Company's refinery. The Florence Oil and Refining Company tapped a large flow of gas in 1892, which it distributed to local households and businesses. At one time, the field's two principal operators provided gas for about a hundred customers.

The price of natural gas was competitive with coal only during the summer months, when rates were reduced by one-third to one-half of the winter schedule. Customers paid, depending on the season, from $1 to $3 a month for hookups of cookstoves and space heaters. There were no meters. In 1899, United cut back on the distribution of gas to conserve dwindling supplies for use under stills in its refinery. Seven years later, all of the field's gas reserves were earmarked for use by the refineries.

Elsewhere in Colorado, natural gas was consumed, as at Florence, in local markets. In the mid-1890s, two exploratory wells near Garcia, in Las Animas County, encountered natural gas. There was no commercial outlet, so the gas was piped to local ranches, where it was used for illumination and heating. In the Boulder field there was one gas well; all others yielded gas as a by-product of oil production.

There were no reliable statistics for natural gas production in Colorado until 1919. That year the output was small, and did not increase significantly until 1924, after large gas fields were discovered in Larimer County, east of the mountains, and in Garfield, Moffat, and Rio Blanco Counties, west of the Continental Divide. The Union Oil Company of California made the state's first discovery of natural gas in commercial quantities on the Wellington dome north of Fort Collins. The discovery well flowed at a rate of 82 million cubic feet of gas, plus about seventy barrels of oil a day.

The Union Oil Company's engineers, after a thorough test of the Wellington well, concluded that it could not produce sufficient gas to supply Denver, but there was enough to serve Cheyenne, Wyoming, and Fort Collins for a number of years. Colorado's capital city had a population of 225,000, compared to 22,000 in Cheyenne. The Cities Service Company, which then owned the Public Service Company of Colorado, Denver's gas and light enterprise, organized a subsidiary, the Colorado-Wyoming Gas Company, to deliver the Wellington dome gas to nearby population centers. A 6-inch line was completed to Fort Collins in 1924, and to Cheyenne in 1926. However, estimates of the supply proved greatly inflated, with the result that the natural gas had to be mixed with coal gas before it was distributed to customers in the two communities. Faced with ongoing decline in the output of gas, the utility had to either find another source of natural gas or revert, in time, to the distribution of the manufactured variety.

Fortunately, at that time a seemingly unlimited source had been located in the Texas Panhandle. Hoping to tap that supply, the Public Service Company of Colorado secured a franchise from the City and County of Denver for the exclusive distribution of natural gas through existing mains. The utility immediately contracted for the interstate transmission of gas to serve additional communities. Within three months, the Colorado Interstate Gas Company, then an affiliate of Standard Oil of New Jersey, commenced construction of a 20-inch pipeline to transport gas from the Panhandle to Denver via Pueblo and Colorado Springs. Gas flowed the length of the 340-mile line on June 23, 1928. Extensions served towns in the Arkansas Valley. One to the area east of Pueblo supplied La Junta, Rocky Ford, and Swink; another to the west served Florence and Cañon City, the principal communities in the prolific Fremont County coal field.

Ironically, the region's principal producer and distributor of coal, and operator of the only integrated iron and steel plant, played a key role in bringing natural gas to eastern Colorado. The Colorado Fuel and Iron Company, then controlled by the Rockefeller family, was asked by the Colorado Interstate Gas Company to substitute natural gas for the so-called producer gas it manufactured from coal for heating ingots in rolling mills. After first declining to buy natural gas, CF&I's management signed a long-term contract in 1927. That was the final encouragement the transmission company needed to build the interstate pipeline.

Some critics assumed that because the Rockefeller family controlled Colorado Interstate Gas, the steel company's decision to utilize natural gas for some operations was an attempt to maximize the transmission firm's profits. This was not the case. By adopting natural gas the Pueblo steelmaker cut fuel costs in its rolling mills approximately in half. At the same time, it eliminated the need for about 2.5 million tons of coal annually at the Pueblo plant. This hastened the decline of coal operations, and dealt a severe blow to a mining industry that already suffered from excessive capacity.

The shift from coal to natural gas was accelerated by the construction of pipelines to additional communities in eastern Colorado. By 1930 the Colorado-Wyoming Gas Company had built lines that carried Texas gas to Boulder, Fort Collins, and other towns ands cities in northern Colorado.

Within a few years, many longtime consumers of coal in eastern Colorado had switched to the clean-burning fluid fossil fuel. In 1929, gas from the Amarillo field was transported to Cheyenne via the pipeline originally laid to carry gas from the Wellington field to the Wyoming capital. In 1930, the Colorado Gas and Utilities Company completed a line that brought gas from the vast Hugoton field of Kansas to Lamar, Holly, Springfield, and other towns in southeastern Colorado for household and industrial use.

Advances in the metallurgy of steel had encouraged the construction and operation of high-pressure pipelines for the safe and economical transmission of natural gas over great distances, as from Texas and Kansas to communities in Colorado. Thin-walled, electric-welded pipes with on-line compressor stations moved large volumes of gas hundreds of miles to markets once filled by manufactured gas or coal. Those fuels could not compete with the high heat value, low cost, convenience, and purity of natural gas. This encouraged exploration and development of more gas fields and the construction of additional pipelines to untapped markets.

The consumption of natural gas in Colorado increased at a fast pace as pipelines linked producing fields with population centers. Where manufactured gas had been in use, the consumption of natural gas carried through established mains required only an adjustment of furnace orifices, or apertures, by means of which gas flowed to the burner. In 1923, Colorado homes and businesses consumed only 800,000 cubic feet of natural gas. Most of this was consumed in oil fields by drilling rigs and pumps. Two years later, following the opening of the field near Fort Collins, consumption jumped to 574.4 million cubic feet. In 1930, when consumption in Colorado rose to 16.6 billion cubic feet, 88.2 percent of that amount was imported from the Texas Panhandle. The remainder was produced within the state, or imported from New Mexico and Kansas. By that date, the manufacture of coal gas had ceased in towns and cities where natural gas was available.

Natural gas was consumed in Colorado in 1930 by households, commercial establishments, and industries. Single- and multiple-family private residences, whether owned or rented, consumed 3.8 billion cubic feet. Commercial hotels, restaurants, wholesale and retail stores, and public and private service enterprises consumed 5.1 billion cubic feet. Electric utilities, plus industries in transportation, mining, manufacturing, construction, agriculture, and forestry, consumed 11.5 billion cubic feet. The state's largest single consumer of natural gas was its principal heavy industry, the Colorado Fuel and Iron Company at Pueblo.

In western Colorado there were few local outlets for the output of the area's numerous gas fields. Residents of Durango gained access to natural gas, in place of the manufactured variety, in 1930. It was imported by the

Mesa Grande Gas Company of New Mexico. The Moffat and Iles fields, in Moffat County, produced gas in conjunction with crude oil. This was also true of the Tow Creek field in Routt County. The Thornburg and Garmesa fields, straddling the Garfield and Mesa county lines, were, at least for a time, exclusively producers of gas. Owners of most gas wells had to shut them down, use the fuel for drilling new wells, waste it by venting, or burn it to produce carbon black. The Rio Blanco Carbon Company operated burning houses on the White River, twenty miles north of Meeker, in the mid-1920s. The combustion of a thousand cubic feet of gas yielded two pounds of carbon black. Each day, about 1 ton was shipped to Denver for use in the manufacture of printing inks and writing fluids.

Gas producers in some of the state's northwestern counties gained access to large urban markets for the first time in 1929. That year the Western Public Service Corporation completed, at a cost of $24 million, an 18-inch pipeline to transport natural gas a distance of more than 300 miles from wells in southwestern Wyoming to customers in Salt Lake City and Ogden, Utah. Western Public was a holding company formed by the Ohio Oil Company, which had discovered the rich Baxter Basin field near Rock Springs; the Producers and Refiners Corporation, a leading producer of oil and gas; and a group of Pittsburgh capitalists. As subsidiary enterprises, the Mountain Fuel Supply Company produced gas or bought it from owners of wells, and the Uinta Pipe Line Company transported gas to the Utah communities, where it was distributed to households and businesses through mains that had once carried manufactured gas. Mountain Fuel Supply completed a line from the Hiawatha field, in northern Moffat County, to a junction with the Wyoming-Utah line in 1929. In the mid-1930s, the Western Public Service Corporation and its subsidiaries were reorganized as one enterprise known after that time as the Mountain Fuel Supply Company.

Colorado's natural-gas industry, after a slow start, expanded rapidly in the 1920s. The Great Depression that began in 1929 slowed exploration and discovery of new fields and undermined what had been an expanding market for gas. Severe economic stagnation lasted until 1934, after which partial recovery occurred until undermined by a recession in 1937–1938. Prosperity eluded the industry until World War II restimulated demand.

5

Coal from Bust to Boom Again 1930–1973

The Colorado coal industry's recession worsened at the end of the 1920s with the onset of the Great Depression. World War II temporarily restored stability to coal mining, but the recession recurred following the close of the conflict. By 1960, recovery was under way as utilities purchased record amounts of coal to generate electricity in an effort to satisfy a fast-rising demand for power from household, business, and industrial consumers.

The 1930s were a bad time for Colorado's coal industry. Output plummeted from 8.2 million tons in 1930 to 5.2 million tons in 1933. The number of employees declined at the same time from 10,683 to 8,179. For those who were fortunate to have jobs, the average days of annual employment dropped from 165 to 140.

Relief from shrinking markets, reduced revenues, unemployment, and underemployment eluded the beleaguered industry until President Franklin D. Roosevelt took office in 1933. At the urging of the chief executive, Congress approved initiatives designed to halt deflation, stimulate recovery, and prevent another economic calamity of similar proportions. A measure of the troubled times was the creation of the National Recovery Administration, which granted hard-pressed industries the opportunity to create "codes of fair competition" designed to stimulate revival of business activity. Mine owners in Colorado, as elsewhere, quickly adopted the Bituminous Coal Code in the hope of raising prices and output.

Under the terms of the code, operators were permitted to establish production quotas, uniform labor standards, and minimum prices, all of which afforded some relief from the unrestrained competition that had undermined economic order in the solid-fuel market for more than two decades. As a quid pro quo, employees gained the right to select collective-bargaining agents. This guaranteed workers a voice in determination of wages and conditions of labor.

When the federal courts challenged the validity of codes in 1935, Congress, convinced that the industry needed ongoing relief, approved new legislation designed to balance supply and demand. The Coal Act of 1935, popularly known as the Guffey-Snyder Act, authorized the formation of a Bituminous Coal Commission and other boards to undertake the tasks of advancing and harmonizing the interests of miners, mine owners, and consumers. Stability was to be derived from output quotas and industry-wide price and wage scales. Workers were again assured rights to organize and to bargain collectively with employers. In May 1936, before the law was fully implemented, the U.S. Supreme Court invalidated its labor provisions and, indirectly, the pricing arrangements.

The Coal Conservation Act of 1937, also known as the Guffey-Vinson Act, revived in slightly modified form the fair-practices provisions of the earlier law. Implementation was slow until the Supreme Court ruled favorably on the price-fixing provisions in 1940. By that time, however, the heightened demand for fuel as the United States mobilized for national defense eliminated the need for curbs on competition within the coal-mining industry.

The legislation of the 1930s provided important gains for Colorado's coal miners and mine owners. Those who extracted coal from the earth gained the right to join the unions of their own choice, and to bargain, free of coercion and duress, for improved wages and conditions of labor. Operators acquired some control over output and prices.

The Colorado industry's darkest days were in 1934, after which new federal programs and a modest economic recovery resulted in gradual improvement. By 1937, 9,543 miners produced 7.2 million tons, an achievement that suggested that coal mining was on the road to recovery. However, grounds for optimism vanished when, in 1938, the state and the nation experienced a recession within a depression. Coal production dropped sharply, as did employment in mines. Slow recovery was again evident by 1939 and continued throughout 1940. In the latter year, however, production remained below the level of 1937.

Contraction, a prominent feature of the 1920s, continued in the 1930s. Numerous mines closed as mine owners sought balance between supply and demand. By 1940, the Colorado Fuel and Iron, Rocky Mountain Fuel, and Victor-American companies had a combined output of 2.2 million tons, one-third of the amount reported in 1930. CF&I operations dwindled to six mines, Rocky Mountain Fuel to two, and Victor-American to four. Oil and natural gas captured an ever larger share of the home, commercial, and industrial markets.

A resounding turnaround occurred in the early 1940s because of abnormal wartime demand for solid fossil fuel. The expansion of war-related industries called for unprecedented energy consumption. Mine operators enjoyed a revival of prosperity, and the UMWA, under the leadership of John L. Lewis, took advantage of the nation's urgent need for fuel to bargain effectively for coal miners' rights to better wages and working conditions. Unpopular but successful strikes resulted in improved benefits, including pensions, as well as welfare and medical programs.

Output jumped from 7 million tons in 1941 to 8.2 million tons in 1944, then declined to 7.7 million tons the following year. The number of active mines dropped from 332 in 1941 to 233 in 1945. Machine mining scored dramatic gains, in part because higher prices and wages were powerful incentives for increasing output while reducing costs. In addition, four years of military mobilization created a severe scarcity of labor.

World War II was followed by a prolonged recession reminiscent of the 1920s. The decline in employment and output continued for more than a decade. By 1950, a force of 4,837 men produced only 4.3 million tons of coal. CF&I closed one mine, the Cameron near Walsenburg, in 1946. The Rocky Mountain Fuel Company closed its last two mines, the Industrial at Superior in Boulder County, in 1945, and the Columbine, near Erie, in 1946. Victor-American had closed its Chandler Mine in Fremont County in 1942 and halted production at the Pinnacle, in Routt County, in 1949.

The industry had to respond to changing patterns of fuel consumption. Railroads, industries, and homeowners abandoned coal for the clean fuels, oil and gas. The continuing decline in coal consumption forced operators to close mines wholesale fashion until production dropped in 1958 to less than 3 million tons. That year, only 2,215 men were employed in and about the state's mines. CF&I ended operations at Crested Butte and at the Nonac Mine, south of Cañon City, in 1952; at the Kebler No. 2, north of Walsenburg, in 1953; and at the Morley, on Raton Pass south of Trinidad, in 1956. Victor-American closed its last two mines, the Wadge in Routt County in 1951, and a year later, the Delagua, northwest of Trinidad.

However, there was one area of stability in the otherwise dreary spectacle of ongoing coal-mining recession. Metallurgical coal enjoyed a steady, even advancing market because of its vital role in coke-making, which was used for the manufacture of pig iron, the raw material for making steel. Historically, three areas in Colorado had supplied coal for coke-making: the Trinidad district of Las Animas County, the Somerset–Carbondale–Crested Butte district of Gunnison and Pitkin Counties, and the Durango district of La Plata County. Because of Durango's remoteness from large markets, its small coking industry collapsed with the demise of the local smelter in 1930. The Purgatoire River Valley, west of Trinidad in Las Animas County, was the principal source of metallurgical coal for the Colorado Fuel and Iron Company, used exclusively in its by-product ovens at Pueblo. The North Fork of the Gunnison River, in western Gunnison County, was an important source of coking coal, but it was too far removed from Pueblo to have any value for coke-making at the steelworks.

Metallurgical coal production increased rapidly in Colorado and elsewhere in the West during World War II. CF&I's blast furnaces operated at near-capacity four years in a row. The by-product ovens produced a record 676,209 tons of fuel in 1942. Output remained at about that level in 1943, before falling to 650,571 tons in 1944 and 639,099 tons in 1945. Battery E, comprising seventy-four ovens with a capacity of 2,800 tons a day, was completed and in operation by 1945. The Defense Plant Corporation built Battery E and leased it to the Pueblo enterprise. CF&I shut down the battery in 1945, but purchased it from the government in 1947, after which it resumed operation.

In addition to the expansion at Pueblo, the federal government encouraged the erection of two new fully integrated iron and steel plants in the West. One was built by Henry J. Kaiser at Fontana, fifty miles east of Los Angeles, to manufacture steel plates for shipyards on the Pacific Coast. The Kaiser plant, comprising blast furnaces, by-product ovens, and rolling mills, produced rolled steel before the close of 1942. Initially the firm purchased coal for its ovens from the Colorado Fuel and Iron Company, and from the Rocky Mountain, St. Louis and Pacific Company, the largest operator in New Mexico's portion of the Raton Basin. Eventually Kaiser acquired its own mines in Colorado, New Mexico, and Utah.

A second integrated iron and steel plant was built seventy miles south of Salt Lake City near Ironton, Utah, where the Columbia Steel Company had manufactured pig iron since 1924. The United States Steel Corporation purchased the Utah plant in 1929 in order to gain access to the lucrative Pacific Coast market through Columbia's steel fabrication plants at Pittsburg and Torrance, California. During the 1930s, Columbia produced about 200,000 tons of pig iron a year. After the United States entered World War II, a second blast furnace was added to the ironworks, and another 500 by-product ovens were built at the Columbia Mine in Carbon County, 130 miles east of the Ironton works.

In 1942, the Defense Plant Corporation selected the Columbia Steel Company to erect an integrated iron and steel plant at Orem, near Ironton. With a rated capacity of 700,000 tons of steel ingots a year, it featured blast furnaces, by-product ovens, open-hearth furnaces, and rolling mills. The federal agency leased the plant for the duration of the war to the United States Steel Corporation, which operated it as the Geneva Steel Company.

By 1945, three western steel plants operated a total of 586 ovens and produced 1.6 million tons of coke a year. The Geneva Works, with 308 ovens, had an output of 731,306 tons, followed by CF&I with 188 ovens and an output of 639,099 tons. At that time, Kaiser was the smallest of the regional steelmakers.

In addition to blast-furnace coke, the by-product ovens of the three companies yielded large amounts of oven gas, which, stripped of chemicals, was consumed by the steelworks. Ammonia, tar, and light oil yielded a variety of products, including ammonia sulfate and benzene, used in the manufacture of styrene, phenol, and aniline, essential ingredients for making synthetic rubber. Other products were toluene, a solvent used in making dyes and explosives, and xylene and naphthalene, essential for making dyes and other compounds.

At the close of World War II, the western steel industry had an uncertain future. The unresolved question was: Could three integrated plants survive and prosper, where only one had been active prior to 1942? Officials of the Colorado and California enterprises insisted that a consolidation of at least two of the firms could ensure success. Henry J. Kaiser wanted to bring all three into a single corporation under his control as an enlarged Kaiser Steel Company. Charles Allen, who organized a syndicate that bought CF&I from the Rockefeller family in 1944, wanted to combine the Colorado and Utah plants. Initially, United States Steel Corporation expressed no interest in acquiring the plant at Orem, but changed its mind when Kaiser and Allen talked of consolidating the western enterprises. In 1946, the Surplus Properties Administration sold the Geneva works to the eastern steel giant for $47 million, one-fourth its original cost. The Geneva Steel Company was absorbed and operated as the United States Steel Corporation's Columbia-Geneva Division.

With three active steel plants in the region, the coke-making industry remained relatively stable for nearly thirty years after World War II. Production and consumption dropped only during periods of recession, of which there were four between 1948 and 1961. Production of steel and coke rose dramatically during the Korean War of 1950–1953 and the war in Vietnam in the 1960s and 1970s.

The Colorado Fuel and Iron Corporation restructured mining operations in the 1950s. Production was restricted to metallurgical fuel for the by-product ovens at Pueblo. At the same time, the Crested Butte, Morley, and Frederick mines were closed because they were old, costly to operate, and without adequate reserves for long-term operations. A new mine was opened to ensure regular supplies of coking coal for the steelworks.

As early as 1945, CF&I's engineers discovered a new source of metallurgical coal near Weston, in the Purgatoire River Valley west of Trinidad. Initial diamond drill tests revealed 25 million tons of coal. Additional investigations revised that figure upward to more than 80 million tons. In 1951, the mining department opened the Allen Mine, driving a slope to gain access to a 5-foot vein. Mining machines stripped the coal from the seam, after which it was moved mechanically to the surface for processing and shipment by unit trains to Pueblo. A wholly owned subsidiary, the Colorado and Wyoming Railway, provided the cars and motive power for a twenty-seven-mile journey from the Allen Mine to Jansen, on the outskirts of Trinidad. At that point, the Denver and Rio Grande Western Railroad hauled the coal another eighty miles to the steelworks.

Unit trains were also active in western Colorado. The United States Steel Corporation purchased the Somerset Mine, in western Gunnison County, in 1955, but did not take over operations until 1961. In addition, it contracted for the output of the Mid-Continent Coal Company's

The Colorado Fuel and Iron Corporation's Allen Mine, near Weston, opened in 1951, was one of the first modern underground mines in Colorado. A high-technology operation, it utilized electric trolleys for moving coal on the surface and underground. Courtesy, Bessemer Historical Society, Pueblo.

mines in Coal Basin, near Redstone in Pitkin County. The output of the Colorado properties was hauled by unit trains to Wellington, in eastern Utah, where it was cleaned and blended with fuels from other districts to serve the needs of the steel company's by-product ovens.

By the 1970s, Colorado's metallurgical fuel industry faced an awesome challenge. Coke-makers had to reduce atmospheric pollution to conform to local, state, and national air-quality standards. Historically, the industry had been known for high pollution. By-product ovens emitted unacceptable levels of fumes and particles. In order to reduce that pollution, western steelmakers had to accept a heavy financial burden.

Having the oldest plant, Colorado Fuel and Iron, renamed CF&I Steel Corporation in 1966, had to upgrade its ovens to comply with clean-air standards. In 1971, it contracted with the Koppers Company for two new batteries of low-polluting ovens, and for the modification of a third to

Unit trains were first used in Colorado to move coal from the Allen Mine to the steelworks in Pueblo. A Colorado & Wyoming train loaded coal by passing through a storage silo at the Allen Mine (1974). The C&W locomotive and crew took the train as far as Trinidad, where a Denver and Rio Grande Western locomotive and crew continued the trip to Pueblo and returned the empty cars to Trinidad. The C&W completed the round trip to the mine near Weston, and loaded the cars for another trip. Courtesy, Bessemer Historical Society, Pueblo.

reduce pollution. Sixty-five new ovens replaced Battery B, which had produced 6.2 million tons of coke in fifty-three years of operation. The new state-of-the-art ovens reduced emissions of gases as well as particles.

A conspicuous feature of CF&I Steel's reconstruction program was an effort to reduce the role of coking ovens in the operations of the Pueblo plant. An electric furnace was installed in 1972 as a replacement for the last of the company's open-hearth furnaces, an outmoded method of making steel. Commencing in 1973, the electric furnace, with a rated capacity of 310,000 tons a year, transformed several tons of scrap metal into steel in two to three hours. The Public Service Company of Colorado built the Comanche coal-fired steam-generating station south of Pueblo to provide electricity. That 350,000-kilowatt plant contracted with AMAX (formerly

American-Metals Climax Inc.) for coal from its Belle Ayre surface mine in Wyoming's Powder River Valley. A Burlington Northern unit train delivered 11,000 tons of coal per trip to the plant, making the 1,200-mile circuit in seventy hours.

Steel without the expense of blast furnaces was an essential step toward cost reduction at a time when American companies found it increasingly difficult to compete with plants in Asia and South America. The high cost of making steel in America hampered competition in a global economy. Costs had to be reduced if the metallurgical industry was to remain viable. For this reason, the future of metallurgical coal mining was more uncertain in 1973 than it had been for at least half a century.

Apart from metallurgical fuel, coal producers in the postwar years could not hold on to their traditional markets. Railroads, by the 1950s, had largely completed the change from coal-fired to diesel-electric locomotives. Industries had converted to oil and gas. What remained was the home market, and it too was moving toward oil and gas.

Householders eagerly changed from coal to clean, convenient fuel oil or natural gas. They welcomed the opportunity to escape the daily chore of shoveling dirty solid fuel into furnaces and removing the large volume of ash that had to be disposed of as trash. Coal bins, a source of dust and dirt, discouraged homeowners from using basements for anything more than laundry or storage. With the adoption of gas and oil, the dreaded task of reheating the house each morning required only the flick of a switch, and combustion did not produce a troublesome by-product. In addition, without the need to store large amounts of coal, basements could be converted into attractive, useful living spaces.

Nationwide, coal claimed the largest share of the home-heating market in 1950. It supplied 33.8 percent, more than any other fuel. Natural gas accounted for 26 percent, fuel oil 22.1 percent, and electricity a meager 0.6 percent. By 1960, coal's share had fallen to 12.2 percent; ten years later it was less than 3 percent. The loss of the home-heating market appeared to be a near-fatal blow to the mining industry.

In Colorado, as elsewhere in the nation, coal-mining enterprises extracted coal mostly from deep deposits. In an effort to cut costs in hope of holding on to existing markets, even expanding into new ones, operators increasingly made underground mining a capital-intensive industry, turning to machines that permitted higher levels of efficiency in the extraction of coal. Electric and compressed-air undercutting machines and drills pre-

Continuous-mining machines were used in the Allen Mine as early as the 1950s. One machine extracted coal from the seam and loaded it into cars or onto mechanical conveyors for the trip from the underground workings to the surface. Courtesy, Bessemer Historical Society, Pueblo.

pared the seam for the explosives used to shoot coal from the working face, while mechanical loaders, shuttle cars, and portable conveyors expedited its movement from the depths of mines to tipples, where it was prepared for shipment.

The most important technical advance in the 1950s was the introduction of the continuous-mining machine, which was invented by Harold Silver, proprietor of Silver Engineering Works in Denver, at the behest of Carson W. Smith, president of Consolidated Coal and Coke Company of Denver. In 1938 Smith approached Silver, an engineer whose talents for invention had been applied, until then, largely for the benefit of the beet-sugar industry. Smith wanted a machine that would make the mining of coal less labor-intensive and, therefore, less costly per ton of output.

By the 1940s, machines had replaced hand labor in all but the smallest mines. Mechanical cutters, drills, and loaders were standard equipment, and the roof-bolting machine was adopted in the following decade. In spite of these advances, the high labor cost of extracting coal undermined the solid fuel's ability to compete with oil and gas.

Smith appealed to Silver's inventive genius in hope of securing a machine that would allow one or two men to perform all of the tasks involved in conventional mining—undercutting, drilling, shot-firing, loading, and hauling coal from the working face to the surface. Silver completed a design for a machine in 1940, after which a prototype was built and, in 1943, tested in Consolidated's Baum Mine at Frederick, in Weld County. By 1946, Silver's engineers had corrected the problems that were inevitable in a new machine of radical design, and it was first used that year for commercial production.

The Joy Manufacturing Company of Pittsburgh, the nation's largest mining-machinery manufacturer, bought exclusive rights to the machine in 1947. Silver shared royalties from sales with Consolidated Coal and Coke for a period of twenty years. Early in 1948, Joy tested the continuous miner in the Mathies Mine of the Pittsburgh Coal Company, and its superiority over all other methods of extracting coal from underground seams led to its widespread adoption in the United States.

The Joy continuous miner, a low-slung, compact machine that measured 6 feet in width and 20 feet in length, mounted on caterpillar treads, created a revolution in coal mining. It cut coal from the face of a seam and moved it to the rear, where it was deposited either in shuttle cars or on a belt conveyor for the journey from the mine. One machine, as depicted in Figure 3, combined in a single coordinated process what formerly had been separate tasks by machines manned by several mine employees. The continuous miner combined the tasks of undercutting, drilling, blasting, and removing the coal from the face of the working seam. Because Silver's invention increased output and reduced labor costs, it dramatically shaped the evolution of coal mining in the second half of the twentieth century.

That remarkable technical achievement, however, did not halt the decline in coal's share of the energy market. By the 1960s, most of the longtime major suppliers of coal to railroads, industries, and domestic consumers had disappeared. In that decade, the onetime giant of the regional fuel trade, the Colorado Fuel and Iron Corporation, changed its name to CF&I Steel Corporation in recognition of the fact that fuel production was for its own use rather than for the regional market.

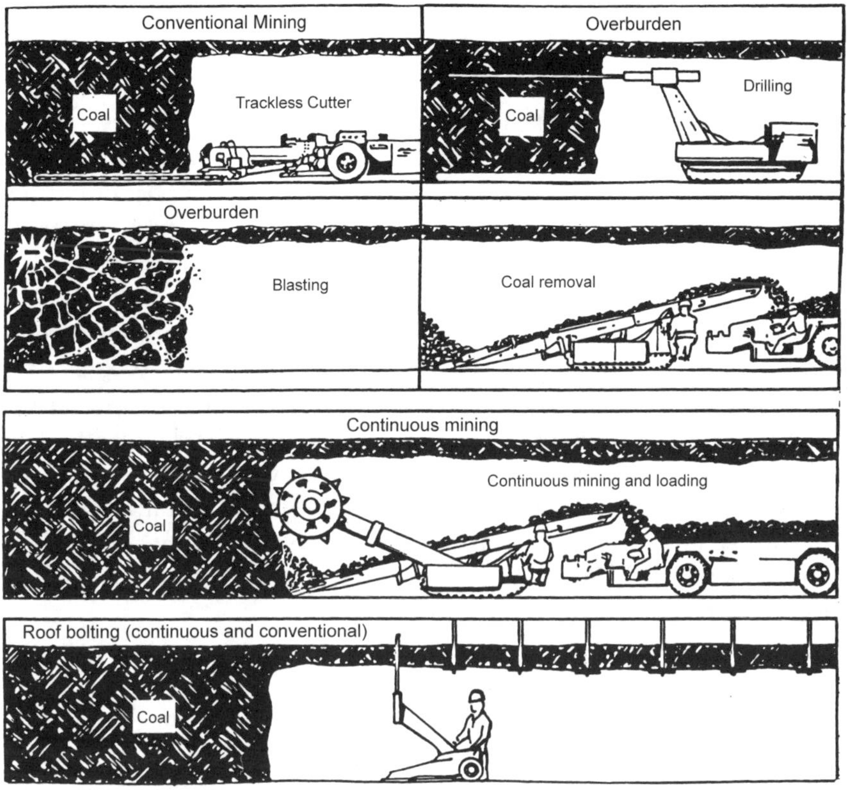

Figure 3. Machines for Mining Coal Underground. From U.S. Office of Technology, *The Direct Use of Coal* (1979).

The contraction of the coal industry created widespread dislocation for those who produced the solid fuel. In 1941 there had been 332 mines with 7,878 employees. Twenty years later, 118 mines provided jobs for only 1,657 men. Miners and their dependents were uprooted from traditional surroundings because companies, if they owned the mining towns, invariably dismantled all structures when they shut down properties. In Fremont County and in the northern counties of Boulder and Weld, company towns were largely unknown, but the closing of mines there did not create less hardship for miners and their families. Men who had worked all their lives in a single industry were forced to seek employment elsewhere, invariably in other occupations. Some relocated to communities with active mines, but most gravitated to the region's industrial centers, where they found employment in manufacturing and service industries.

With the shrinkage of the coal market, individuals and corporations active in the mining industry, as well as politicians from coal-producing states, insisted something had to be done to reverse that trend. Congressmen John Saylor of Pennsylvania and Wayne Aspinall of Colorado were the principal advocates of new research efforts to expand markets for coal. They wanted to establish an independent coal research and development commission, but President Dwight Eisenhower insisted that new programs for coal research had to be housed, along with the Bureau of Mines, in the Department of the Interior. The Coal Research and Development Act of 1960 created the Office of Coal Research, which was empowered to make research grants to public and private enterprises, including universities, for the purpose of improving the mining, preparation, marketing, and utilization of coal in the United States. In the long run many projects funded by the agency had a positive impact on the industry.

Ironically, a new boom began even as the industry's bust was altering the demographics of Colorado's coal-producing counties. By 1960, rising demand for the solid fuel halted the long decline. The state's mines reported a total yield of 11.2 million tons, the first increase in several years, and the upward trend in output continued throughout the decade. The revival was triggered by demand for fuel from steam-generating electric plants. Indicative of change was the Public Service Company's new power plant at the Cameo Mine in Mesa County. Completed in 1957, it was the state's first modern "mine-mouth" plant. Coal moved from the mine portal directly to the electric plant over belt conveyors. The construction of generating stations at Nucla, Hayden, and Denver, and the expansion of older facilities, sustained the growth of coal production throughout the 1960s and into the following decade.

The resurgence of coal production was accompanied by the rapid development of surface, or what were popularly called open-pit, mines. The first of these mines had opened in 1866 near Danville, Illinois, where the overburden was removed with horse-drawn plows and scrapers, and the coal loaded by hand into wheelbarrows and wagons. Production from open pits evolved slowly for nearly a century before expanding at a lively pace during and following World War II. Surface mines accounted for 1.5 percent of the nation's coal production in 1920, 4.3 percent in 1930, 9.4 percent in 1940, and 23.9 percent in 1950. Improvements in equipment and extraction techniques, plus lower production costs, encouraged rapid growth in the number and size of surface mines wherever there were shallow deposits of bituminous and subbituminous coal.

In the 1960s, large production mines in the intermountain West were nearly always surface units. By 1970, nearly two-thirds of all coal produced in the intermountain states came from these operations. This was largely because coal was in great demand as fuel for generating electricity. New plants were built, and old ones expanded, to satisfy the rising demand for electric power in the region, as well as in the energy-starved cities of the Southwest and Pacific Coast. In addition, western coal, because of its low sulfur content, became the fuel of preference for many utilities as the nation struggled to control air and water pollution as the result of the ever increasing industrial consumption of fossil fuels.

The expansion of coal mining throughout the intermountain West, particularly in Montana and Wyoming, created widespread uneasiness about the potentially harmful effects of large-scale energy production on the environment. Indicative of this concern was the public's reaction to the "North Central Power Study," issued by the Federal Bureau of Reclamation in October 1971. Responding to a growing shortage of energy, the study identified an immediate need for more than three dozen coal-fired steam-generating plants with an aggregate capacity of 10,000 megawatts in Montana, Wyoming, and the Dakotas. This created fear that large portions of the Great Plains, including the coal-rich Powder River Basin, would be dug up by draglines and power shovels, and communities overrun by raucous newcomers, so that the cities outside of the Rocky Mountain region could have ample supplies of electricity. Disturbances attributable to the production, marketing, and consumption of coal had, environmentalists noted, both short- and long-term impacts that affected in some way the quality of the region's air, land, and water. Of primary concern was the possibility that by the time the negative impacts of coal-related activities were readily apparent to a majority of Americans, it could be too late to reverse the damage.

For that reason, concerned citizens formed the Northern Plains Resource Council to track and, if possible, discourage the development anticipated by the "North Central Power Study." The council, which drew its support largely from the League of Women Voters of Montana and

South Dakota, the Montana League of Conservation Voters, the Montana Wilderness Association, the National Wildlife Federation, and the Sierra Club, filed suit in Washington, D.C., in 1973, insisting that development of coal resources on the northern plains should be suspended until comprehensive environmental-impact statements had been filed by appropriate federal agencies for each proposed mine and related facilities. In addition, environmental action groups demanded state legislation to safeguard the environment. Remedial legislation was adopted in Montana, Wyoming, and other western states. Agreements with the U.S. Department of the Interior eventually gained for states control over strip mining within their respective borders.

The demand for western fuel set off a rush for leases on federal lands in the intermountain West underlaid with coal. During the years 1965–1971, mining enterprises, actual and prospective, secured leases at bargain rates on 433,000 acres with commercially valuable coal, most of it in the form of strippable deposits. This was one-third more coal land than had been leased by the federal government in the forty-four years following the enactment of the Mineral Lands Leasing Act in 1920. That law stipulated that lessees had to pay a royalty of not less than 5 cents for each ton of coal extracted. The return to the government, less than 1 percent of the market value of the resource, enabled lessees to acquire for a pittance extremely valuable deposits. This provoked an outcry that royalties should reflect the actual value of the resource. In 1971, responding to that concern, the Department of the Interior declared a moratorium on leasing until Congress could act to curb what was widely viewed as the reckless exploitation of the public domain.

At the same time, a growing shortage of energy pointed to the need to expand coal production. It was apparent that domestic petroleum production could not satisfy the nation's accelerating demand for oil. In fact, imports had been rising from year to year for more than a decade, making the American economy vulnerable to interruptions of supplies from abroad. Petroleum industry analysts cautioned that oil imports, amounting to 26 percent of consumption in 1970, could spiral to as much as 60 percent in another decade.

One way to avoid that possibility was to increase consumption of coal, which the United States possessed in great abundance. Coal could satisfy the growing demand for energy, which was expected to approximately double over a period of a decade and a half, rising from 68 quadrillion

British thermal units in 1970 to at least 112.5 quadrillion and possibly as much as 130 quadrillion Btu in 1985. However, increased usage of solid fossil fuel posed a dilemma, for the combustion of coal by industrial plants, including utilities, meant more pollution.

This potential risk motivated Congress to approve a series of bills designed to protect air quality in the United States. Already on the statute books were laws adopted in the years 1963 to 1965 that authorized the federal government and the states to take limited actions to protect the environment, including air quality. In 1967 Congress instructed the Department of Health, Education and Welfare to tighten federal guidelines but left enforcement of air standards to the states.

The National Environmental Policy Act of 1969 and the Clean Air Act of 1970 introduced significant changes. New regulations substituted federal for what had been state standards, and imposed more stringent antipollution requirements for industries and automobiles. Enforcement remained with the states, but a new office, the Environmental Protection Agency (EPA), was to ensure safeguards for the environment as the nation increased consumption of fossil fuels.

Corrective action was initiated by the EPA in 1971, when it set sulfur dioxide emissions standards for newly constructed coal-burning electric plants, the first of a series of actions designed to cope with problems related to the rapid expansion of generating stations. The burning of fossil fuels polluted the atmosphere. Coal contributed by far the most, and natural gas the least, amount of pollutants. Combustion released sulfur dioxide (SO_2), nitrogen oxides (NO_x), and ash in the form of particles. In the atmosphere, SO_2 and NO_x combined with moisture to fall to earth as acid rain, which was harmful to fauna and flora. Sulfur dioxide could be controlled by flue-gas desulfurization, a process in which wet scrubbers sprayed water and lime, which combined with SO_2 to form a sludge, causing it to drop out of the flow of gases in the stack. Nitrogen oxides could be controlled by reducing the temperature of combustion. Particles could be largely removed by means of either electrostatic precipitators or by fabric filters.

Growth of electric-generating capacity reflected rising demand for electricity, which advanced nationally at a rate of about 5 percent a year. Operators of new power stations had the option of burning high- or low-sulfur coal. If they chose the former, which came mostly from mines of the Midwest and East, they had to install flue-gas desulfurization equipment, better known as stack scrubbers, to reduce emissions of harmful materials.

This equipment, which was expensive to buy and to operate, was unnecessary if generating plants consumed low-sulfur coal, the principal source of which was the Rocky Mountain states. The result was the rapid development of western mines in the 1970s.

A growing energy shortage and the rising concern for the environment turned the mining industry's attention to surface-mining operations, which were capable of quickly increasing the output of low-sulfur coal. In the intermountain states there were abundant reserves of the low-polluting coal and lignite at shallow depths, often in seams of 70 to 100 feet in thickness. Since more than half of the nation's strippable deposits were in the Rocky Mountain states, with the largest easily accessible reserves in Montana and Wyoming, the energy boom of the 1970s had an enormous impact on these states. However, other coal-producing states in the intermountain West experienced expanded surface-mining operations.

Strip mines had obvious advantages over underground mines. The output per man per day in surface operations was substantially more than in underground mines, and the cost of coal from the former was on average one-third less than from the latter. The high level of productivity had to be credited, in part, to application of heavy equipment and sophisticated technology, as shown in Figure 4. Operators opted for ever larger and increasingly efficient earthmoving equipment in the form of draglines, shovels, scrapers, and trucks. For example, 140-cubic-yard shovels were in use by the mid-1960s, at a time when 200-cubic-yard models were already undergoing tests. Trucks evolved in a similar progression. Many with capacities of 120 tons were in use in 1964, by which time models with twice that carrying capacity had been designed. A similar evolution to larger sizes characterized rotary drills used to prepare holes for explosives to loosen overburden and exposed coal seams. Front-end loaders and hydraulic excavators placed coal in trucks for movement to preparation plants where it was cleaned, mixed if necessary to obtain a uniform Btu content, and stored until needed either for local consumption, in the case of mine-mouth generating facilities, or for shipment to distant consumers.

The movement of coal from mine to distant markets was invariably by unit trains. That revolution in transportation began in 1962 when railroads responded to the competitive threat of coal-slurry pipelines by experimenting with more efficient methods of haulage and improved equipment for moving coal from where it was produced to where it was to be consumed. The "unit" portion of the trains was generally owned by the

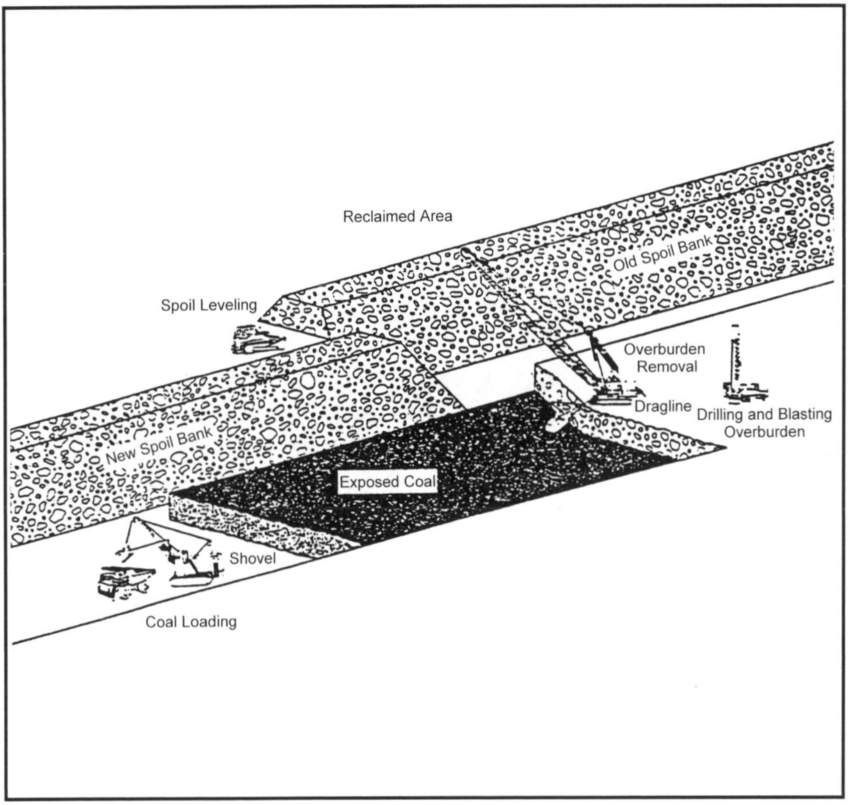

Figure 4. Surface Mining With Dragline and Shovel. From U.S. Energy Information Administration, *Coal Data: A Reference* (1995).

utility company to which the coal was being shipped. The cars were loaded, hauled directly to the generating plant, dumped, and returned to the mine mouth in the shortest possible time. Utilities, by leasing locomotives and track, hastened the delivery and reduced the cost of transporting low-sulfur coal to distant plants. A hundred or more quick-dump cars permitted fast turnaround times from mine to market and back to the mine.

Unit trains delivered enormous tonnages of coal to distant points at nominal cost. In 1967, when one-third of all coal moved to market on unit trains, one rail company, as an experiment, hauled 500 loaded coal cars in a single unit a distance of 157 miles. The train stretched, from end to end, about four miles, and moved a total weight of 47,000 tons, a record at the

A vital link between the mine and the consumer was the local coal merchant, who ordered fuel in carload lots, and moved it by truck, one or more tons at a time, to householders and businesses. Until the 1940s, most Coloradans used the solid fossil fuel for heating. The W. A. Anderson Coal Company was one of many fuel merchants in Pueblo. Courtesy, Bessemer Historical Society, Pueblo.

time for bulk shipments by rail. Eventually, unit trains were standardized at 100 to 125 cars, each holding 100 tons.

In order to lessen pollution, mining companies cleaned the coal before it was sold to utilities. Careful methods of preparation eliminated some, but not all, of the undesirable impurities. Most of the sulfur that remained could be eliminated at the time of combustion by means of flue-stack scrubbers and other devices. However, there were no feasible techniques at the mine or during combustion for coping with carbon dioxide emissions.

One way to burn coal and at the same time reduce negative side effects was to convert it into largely pollution-free oil or gas. This option was widely recognized by the 1970s, when a growing shortage of domestic natural gas and petroleum suggested that coal liquefaction and gasification could ensure ample supplies of fluid and gaseous fossil fuels for the future. The intermountain West, with the nation's largest reserves of high-quality oil shale and strippable low-sulfur coal, was a likely center for the production of what was called synthetic oil and gas. In 1971, with the allocation of increased funds to the Office of Coal Research, Congress authorized incentives for increasing the domestic production of energy. These included large-scale private and public investments in oil-shale coal gasification and liquefaction, the development of geothermal resources, and electric generating plants. Obviously it would take time as well as money to develop the technology essential for the commercial production of alternative and synthetic fuels. The one resource abundantly available was coal, the output of which was expected to grow at a rapid pace in an effort to narrow, if not close, the impending energy gap.

In the short run, increased utilization of coal was invariably linked to the generation of electric power. By 1973, the output of coal in the intermountain states for consumption by electric utilities exceeded 75 million tons, an increase of more than 30 million tons since 1970. At the same time, the West's share of total coal production in the United States jumped from 7.6 to 12.8 percent. Utilities throughout much of the country were buying large amounts of so-called compliance fuel from western mines. The anticipated shortages of oil and gas suggested that the boom could last for a long time.

In Colorado, there was little concern about the negative environmental impact of coal mining in the 1970s, perhaps because the state lacked the enormous resources of Wyoming and Montana for strip mining. Coal had been produced from open pits in Jackson County as early as 1911, and a small amount was mined in a similar fashion in Elbert County from 1921 to 1942. Seven surface mines in 1951 contributed only 9 percent of the coal mined in the state, and two mines in Routt County accounted for most of that production. By 1961, the number of surface mines remained at seven, but together they accounted for 14 percent of the state's output. By 1973, ten surface mines produced 46 percent of the state's output, but employed only 26.8 percent of the workforce in coal mines.

The bulk of the state's accessible strippable reserves were west of the Continental Divide. The first to be opened on a large scale were in Routt County, perhaps best known today as the location of the Steamboat Springs ski resort. Near Oak Creek, twenty-five miles south of Steamboat Springs, the Edna Coal Company extracted coal from surface excavations as early as 1945. That enterprise sold its property in 1961to the Pittsburg and Midway Coal Company, a subsidiary of Spencer Chemical Company of Kansas

City, Missouri. The Edna Coal Company also owned the Navajo Strip Mine in Montrose County. This was sold to the Peabody Coal Company and renamed the Nucla Strip Mine in 1964, when the new owner took over the contract for supplying the local power plant.

By that date, the largest producer in Routt County was Pittsburg and Midway, later a subsidiary of Gulf Oil Corporation. It extracted 461,304 tons of coal from state and federal leases, as well as from private land, and gradually increased production from year to year until 1973, when output exceeded 1 million tons. Ninety-nine percent of the coal was consumed in Colorado, most of it by electric utilities. At that time, the mine had a projected closing date of 1991, when the reserves were expected to be exhausted.

The Energy Coal Company, a subsidiary of Morgan Coal Company of Indianapolis, Indiana, opened a surface mine north of Oak Creek in 1962. It produced about 500,000 tons of coal a year until 1972, when it was sold to Energy Fuels Corporation, which was controlled by Robert W. Adams, whose previous mining experience had been in uranium. Under the new owner, the enterprise opened an additional mine in 1972, at which time it was operating on two federal leases and had applications pending for more. Production amounted to 888,709 tons in 1973. Almost all of the output was sent via a unit train to the plants of the Public Service Company of Colorado in the Denver metropolitan area. Some coal was sold to utilities in the Midwest.

Peabody Coal, in a joint venture with Western Utilities Coal Company, developed the Seneca open-pit mine east of Hayden in 1964 to supply fuel for the first unit of the Colorado-Ute Electric Association's electric generating station at that place. Seneca Strip Mine No. 1, located north of highway U.S. 40, was abandoned in 1968–1969, at which time a second mine was opened south of the highway on private land and a federal lease. Peabody's Seneca No. 2 produced 630,377 tons in 1973.

Colorado's largest consumers of coal were two giant electric utilities, the Public Service Company of Colorado, with headquarters in Denver, and the Colorado-Ute Electric Association, with its main office in Montrose. The Public Service Company operated four power plants in the Denver metropolitan area by the 1970s. In that decade, it constructed the Comanche plant south of Pueblo. The company obtained coal from several sources, mostly by unit trains from Wyoming's Powder River Valley and from surface mines in Routt County.

The Colorado-Ute Electric Association, a producer and distributor of electricity to Rural Electrification Administration cooperatives in Colorado, southern Wyoming, and eastern Utah, opened its first power plant at Nucla in 1959. Four years later it began construction of a $30.5 million, 250-megawatt generating station at Hayden in Routt County. When Hayden No. 1 began operation in 1966, the plant burned 500,000 tons of coal a year from local mines. Hayden No. 2 was completed in the mid-1970s.

In Moffat County, coal production had been principally for local domestic and commercial consumption until the 1970s. Utah International, a subsidiary of General Electric, recognized as early as 1954 the county's great potential for development. Utah International acquired a tract of land with substantial shallow coal deposits south of Craig but did not attempt to open a mine until March 1973, when it signed an agreement to supply 68 million tons of coal over a period of thirty-five years to the Colorado-Ute Electric Association.

Other out-of-state enterprises acquired Moffat County coal lands. The Empire Coal Company, an Oklahoma corporation, purchased the Silengo Coal Company's Wise Hill Mine in 1973. The W. R. Grace Company bought the Colowyo Coal Company's Red Wing Mine, a longtime producing property. The newcomers planned to mine coal southwest of Craig to supply large utilities, including the local Colorado-Ute Electric Association's plants.

By the early 1970s, the coal-mining industry had solved its long-running problem of excessive capacity. Production nearly doubled between 1961 and 1973, and the number of mines declined to thirty-six, less than one-third of those active a dozen years earlier. The number of people employed in the mines remained steady at about 1,500, indicating a substantial rise in the productivity. This reflected the adoption of new technology to extract coal underground, and the opening of additional surface mines. The high cost of mechanization in underground mines allowed the larger enterprises to squeeze out smaller producers, while the capital expenditure required for opening and operating surface mines, and their potential for large output, gave them a cost advantage over other methods of mining.

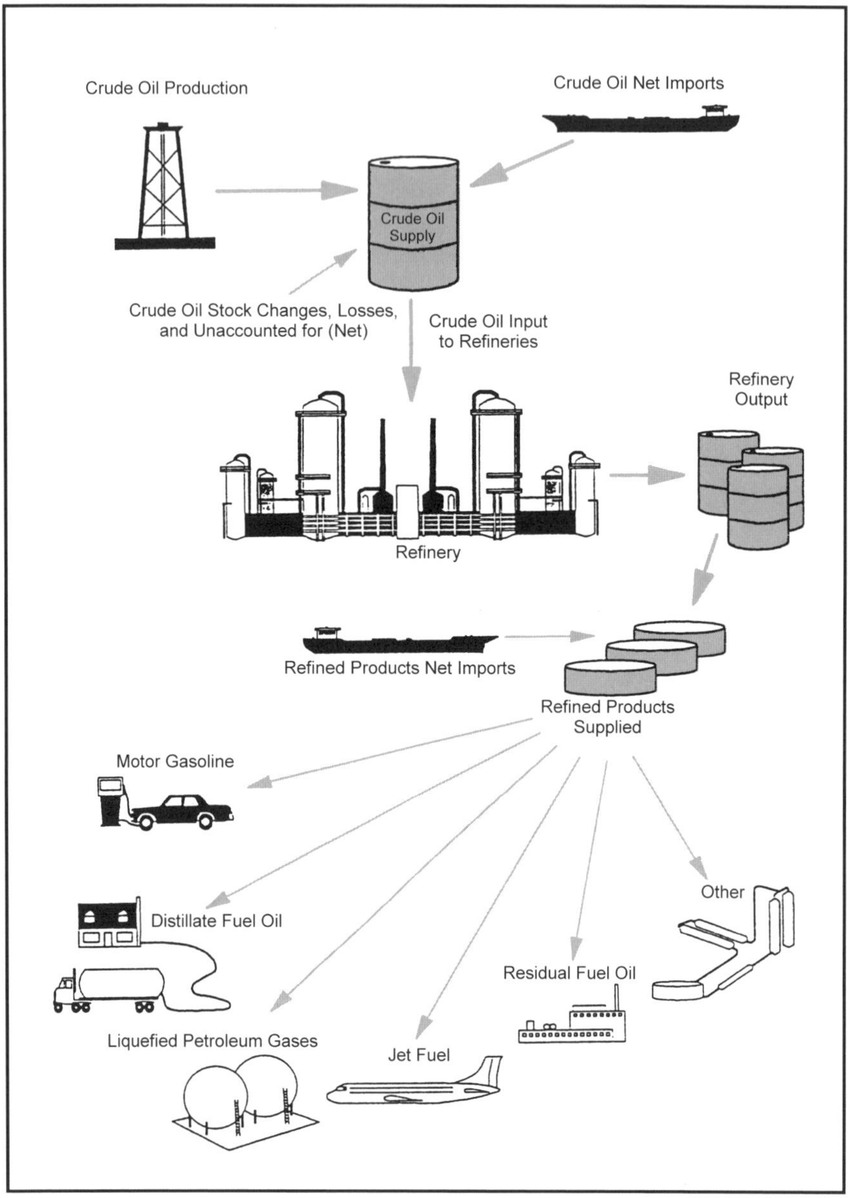

Figure 5. Petroleum Supply and Disposition in the United States. Adapted from U.S. Energy Information Administration, *Petroleum Supply Annual* (1997).

6

Fuel for the Car Culture 1930–1973

The Depression years of the 1930s were a difficult time for the oil industry. In Colorado, as elsewhere in the United States, oil was a glut on the market as early as 1931. That year it sold in some states for as little as 10 cents a barrel. The cause was too much supply and not enough demand. It was a long-term problem in that new fields, particularly in California, Oklahoma, and West Texas, had created surplus stocks of oil during the 1920s. Excess capacity assumed new proportions in October 1930, with the opening of the four-county East Texas field. Within ten months its yield was about a million barrels of oil a day. This, coupled with continued expansion in other states, sharply depressed the price of crude and refined products.

Conditions became so chaotic that major oil-producing states resorted to extraordinary measures in an effort to curb output and to raise prices. The governors of Texas and Oklahoma imposed limits on production and mobilized National Guard units to enforce them. In other states, operators frequently reduced output rather than sell oil at less than the cost of producing it. In Colorado, exploration and development slowed almost to a standstill. One hundred independent well owners at Florence closed down operations when the price of crude dropped to 32 cents a barrel in the summer of 1931. However, the Continental Oil Company continued to pump sixty wells to supply its local refinery.

When states were unable to bring production of crude oil in line with demand, operators looked to the federal government for corrective action. A series of presidential decrees in July 1933 discouraged interstate transportation of oil, and was followed in August by a Petroleum Industry Code promulgated by the National Recovery Administration. The objective was to raise prices by restricting output to approximately the level of demand. Production for the United States was initially set at 24 million barrels per day. Daily production limits were applied to states. Texas led all others with 975,200 barrels, followed by Oklahoma (540,000), California (480,000), Kansas (111,000), the Appalachian states (94,200), and the Rocky Mountain region (80,300), half of which was assigned to New Mexico. Proration—the setting of quotas for producing fields—was left to the states.

In order to strengthen the effectiveness of state laws, Section 9(c) of the National Industrial Recovery Act (NIRA), which created the National Recovery Administration, prohibited transportation in interstate commerce of "hot oil," which was the popular term for petroleum produced in violation of state proration decrees. This feature of NIRA was struck down by the Supreme Court in January 1935, jeopardizing the code's other provisions. Congress responded by approving the Connally Act, which required interstate transporters of crude and refined oil to obtain certificates indicating compliance with state laws.

With the lapse of the Petroleum Industry Code, Congress granted to states authority to enter into compacts to deal collectively with problems of conservation and production. Along with major oil-producing states—Texas, Oklahoma, Kansas, New Mexico, and Illinois—Colorado joined the compact. Nonmember states sent unofficial representatives to meetings. Interstate cooperation, together with voluntary action within the industry, substantially redressed the imbalance between supply and demand during the remainder of the 1930s. By 1941, eleven states were officially active participants in the interstate oil compact.

The Depression stifled growth in Colorado. One new field attracted much attention because of its location on the plains, many miles east of the foothills. The Platte Valley Petroleum Company struck oil at a depth of 6,667 feet, opening the Greasewood pool in October 1930. The well averaged 350 barrels daily for almost a year. The possibility of other pools in the Denver Basin of northeastern Colorado set off a leasing campaign in which approximately 1 million acres were acquired for exploration. Enthusiasm was dampened when only three producing wells were completed in the Greasewood pool. These were overshadowed by six dry holes, which defined the geographical limits of production.

The deepening depression stifled, at least temporarily, exploration and development. The state's production of 1.7 million barrels in 1930 fell to 919,000 barrels in 1933, before rebounding gradually over the remainder of the decade. In the depths of the economic crisis, only three wells were completed throughout the state in 1933. One, a 1,000-barrel well, expanded the boundaries of the Iles dome field in Moffat County. Also of importance that year was a well drilled to productive sand at a depth of 6,300 feet by Standard Oil of California in the Rangely field of Rio Blanco County. Prior to that date, 107 shallow wells, mostly in Mancos shale, had been completed over a period of more than thirty years. The deep well remained shut in for ten years because there was no outlet for the oil. In time, Rangely was to be one of the West's most prolific oil fields.

For the balance of the 1930s, statewide exploration led to oil discoveries almost exclusively in western Colorado. In 1934, the Mountain Supply Company found oil at a depth of 2,345 feet on the Hiawatha dome in Moffat County. Since 1927, the field had produced large quantities of natural gas, sometimes with evidence of oil. The first well to yield oil, rather than gas, marked the emergence of the field as a large producer of crude petroleum. Moffat County gained additional prominence as an oil producer in 1936 when the Mountain Fuel and Supply Company completed a discovery well on the Powder Wash structure.

In 1935, the Price-Gramps pool, in Archuleta County, was discovered southeast of Pagosa Springs, close to the Colorado–New Mexico border. The initial well, completed by an Oklahoma drilling company for the William E. Hughes estate, encountered oil at a depth of 1,120 feet. It had an output of 40 to 50 barrels a day. A second well in the same locale produced from a depth of 1,172 feet at a rate of 500 barrels a day. Lafayette M. Hughes, a prominent Denver attorney, represented the estate, which owned 65,000 acres on what had been the Tierra Amarillo land grant.

In December 1937, the Texas Company, in a joint venture with the California Company, a subsidiary of Standard Oil of California, completed a discovery well on the Wilson Creek structure in Rio Blanco County. Located forty miles southwest of Craig, the well had an estimated output of 2,000 barrels a day. Oil was hauled by truck to Craig, the shipping point by rail and the site of a Texas Company refinery. In 1940, a pipeline was completed from the field to a connection with the line that linked the Iles field with Craig.

In contrast to activity on the Western Slope, intense exploration failed to open new fields in eastern Colorado. Following the opening of the Greasewood pool in 1930, much of the area was leased, reportedly 6 million acres by 1935, and numerous companies—Continental, Gypsy, Indian Territory Illuminating Oil, Ohio, Standard of California, and Magnolia—some of the largest concerns of the day, probed for oil without success. To no avail, the Gypsy Oil Company explored 1 million acres, located between Limon, the seat of government for Lincoln County, and the Colorado-Kansas line. This track, leased from the Union Pacific, was part of the railroad's land-grant properties.

A contemporary theory attributed the lack of new discoveries in eastern Colorado to the extreme depths of oil sands. Acting on this possibility, in the summer of 1936, the Indian Territory Illuminating Oil Company sank wells to depths of 5,500 and 6,500 feet in Washington County without penetrating oil-laden horizons. Geophysical tests indicated the presence of oil in the region, but it eluded drilling crews. A year later, Gulf Oil Corporation completed a duster to a depth of nearly 7,000 feet in Lincoln County. Similar experiences were repeated elsewhere on the eastern plains.

World War II opened new opportunities for petroleum industries. Rising demand for oil forced the federal government and the states to set aside restrictions on the output of crude oil and refinery operations. In May 1942, President Franklin D. Roosevelt issued an executive order that created the Office of Petroleum Coordinator, who assumed the task of designating the quality and quantity of oil to be produced in each state. As petroleum coordinator, Secretary of the Interior Harold Ickes established the Petroleum Industry Council for National Defense, composed of experienced oilmen who assisted the federal government's efforts to mobilize petroleum resources for the conflict with the Axis Powers.

Congress encouraged increased production with the passage of the O'Mahoney Act, signed into law on December 24, 1941. It was designed to stimulate exploration and development on public lands by substituting a flat royalty of 12.5 percent on production from new wells in place of the step-scale rates that, since 1935, had allowed price variations on oil from public leases. The act, introduced by Senator Joseph O'Mahoney of Wyoming, stimulated exploration and development of oil fields in the West, where public lands were available for lease from the federal government.

During the war years, from 1941 to 1945, the annual production of crude oil increased in the United States. Most of this gain was in Texas, California, and Oklahoma. The oil states of the Rocky Mountains advanced their combined output 10 percent, in spite of a 6 percent decline in New Mexico's yield. In Colorado, the output of oil jumped from 3.3 million barrels in 1941 to 5 million barrels in 1945. Most of the increase came from two Western Slope fields, Wilson Creek and Rangely.

Only one new Colorado source of oil was opened during the war years. The Amerada Petroleum Company, following a seismograph investigation, drilled a discovery hole in the Clark's Lake field in Larimer County. Five wells, completed to depths of about 6,000 feet in 1943 and 1944, produced 64,415 barrels in the latter year. The output surged to 94,696 barrels in 1945, before gradually declining to 29,828 barrels in 1949.

Wartime constraints on the consumption of oil and petroleum products ended soon after the close of hostilities. On August 15, 1945, one day after Japan capitulated, President Harry Truman terminated the rationing of gasoline and fuel oils. Within two weeks, most restrictions on production and marketing activities were canceled. Truman officially closed down the Petroleum Administration for War as of May 8, 1946. By that date, the oil industry, responding to rising demand, was poised for another round of rapid expansion.

Colorado gained national prominence with the surging output of the Rangely field. In a manner reminiscent of single-area impact that shaped development in other states, Colorado's production centered for many years in the Rangely district. Intensive production from deep sands began in 1943, spurred by the threat of an oil shortage during World War II, and continued for several years after the close of the conflict. For a time, Rangely was the largest field in Colorado. From a single deep well in 1943, the number grew to 485 wells in 1949. Production peaked in 1956, and then gradually declined.

Initially, there was only a limited market for the Rangely field's oil. Prior to September 1945, crude was shipped by truck 107 miles to Craig, or 210 miles to Salt Lake City. Then the Utah Pipeline Company (a Standard Oil of Indiana subsidiary) completed a 10-inch line from the field to Hamster, Wyoming, where it connected with lines that ran westward from Wyoming's fields to Salt Lake City, and eastward to Denver. Standard Oil of California (Chevron) completed a new 10-inch line to Utah in 1949 to supply its refinery in north Salt Lake City.

The leading operator at Rangely was Standard Oil of California. Other large producers were Stanolind Oil and Gas (Standard Oil of Indiana), Texaco, Continental Oil, Phillips Petroleum, as well as several independents. In 1954, the California Company and Stanolind erected a casing head plant to extract gasoline from natural gas, which was produced in conjunction with the crude oil. Most of the processed gas was reinjected into oil wells as a means of maintaining pressure in the underground reservoir.

Secondary recovery techniques were widely adopted once the field passed peak production. Water flooding was introduced on a limited basis in 1957, the year that Chevron became the field's unit manager. Full-scale water flooding, phased in over a period of five years, stimulated additional output and extended the life of the field for perhaps decades.

The decline of production at Rangely was offset by new discoveries in the Denver Basin, a vast area that stretched across northeastern Colorado into Nebraska and Wyoming. At the time the Boulder field had been opened in 1901–1902, many oilmen, devotees of the "bobber" for locating pools of oil, had insisted that Colorado's high plains were devoid of petroleum, except for a narrow corridor immediately east of the foothills, defined by Ferdinand V. Hayden's investigation for the U.S. Geological Survey thirty years earlier. A dissenting view held that northeastern Colorado contained vast deposits of crude petroleum and natural gas at great depths.

The principal spokesman for this group was Arthur Lakes, onetime professor of geology at the Colorado School of Mines, successful consulting geologist, author of numerous books and articles, and editor of prominent trade journals devoted to mining in Colorado and the West. Based on a study of geological strata upturned by the formation of the foothills and the mountains, Lakes suggested that oil-rich sands were to be found at depths throughout the Denver Basin. The Geological Survey endorsed this view. Lakes acknowledged, however, that because of the limitations of existing technology, it probably was not cost-effective to seek the deep deposits at that time. Widespread adoption of rotary drills and improved techniques for locating underground pools encouraged exploration in the basin in the 1920s and 1930s, but extensive development was delayed until after World War II.

As of 1945, eastern Colorado had been largely neglected by oilmen. Six fields, most of them long past flush production, contributed little to the state's output of oil. The Fremont County field was the oldest, followed in chronological order by Boulder, Fort Collins, Berthoud, Greasewood, and Clark Lake. The output was meager; the combined annual production did not equal 250,000 barrels of petroleum.

There had not been a major discovery in the huge Denver Basin since the Greasewood field, near Orchard in Weld County, in 1930. Only the small Clark's Lake field, north of Fort Collins, was opened in 1944. However, with the end of World War II large leasing programs occurred, for northeastern Colorado was generally recognized as the next likely location for large-scale oil development.

The first successful postwar wildcat on the plains east of the foothills was completed in May 1949, ten miles northeast of Sidney, Nebraska, in what came to be known as the Gurley field. The Ohio Oil Company's Mary Egging No. 1, with a production of high-grade crude at a rate of more than 400 barrels a day, touched off a frenzy of exploration in three states. Several major companies—Cities Service, Stanolind, and Carter (Exxon)—plus numerous independents participated in the rush to secure leases. An estimated 14 million acres were leased in 1950. New fields opened in rapid succession in eastern Wyoming, western Nebraska, and Colorado's Adams, Logan, Morgan, Washington, and Weld Counties. Pipelines linking the basin to markets in the Midwest, as well as to refineries in Colorado, accelerated the pace of development.

One of the largest independents, the Plains Exploration Company, was controlled by Allan W. Biggerstaff and associates of Sterling, Colorado. Biggerstaff, the longtime editor of the *Sterling Advocate*, came to be known as "Mr. Oil," and was recognized as a major figure in the development of the Denver Basin. The Plains Exploration Company drilled several dry holes in the vicinity of Sterling prior to the Ohio company's initial success in Nebraska. Following that discovery in the spring of 1949, the Plains Exploration Company added to its leases and entered into a joint venture with the British American Oil Company to drill the Segelke well in Logan County. The success of that venture revealed a southwesterly extension of deep oil-bearing sands from Nebraska into Colorado, and further stimulated exploration in Colorado's northeastern counties.

Following the success of the Segelke well, northeastern Colorado was the state's most active area of oil exploration. That activity boosted the state's output to all-time records in 1956 and 1957. From 1951 to 1961, the region's fifteen principal pools produced an aggregate of 109 million barrels of oil. The leader was the Adena field, followed by the Little Beaver-East and Yenter fields. Arthur Lakes's prediction of large-scale oil development at depth in the northeastern counties finally had been proved correct.

Wells in the Denver Basin were relatively easy to drill, less costly than in other places, particularly in the mountain district, but the so-called fields were often only pools, which were small, difficult to locate, and short-lived producers of crude oil. The Divide field, for example, was an oil producer for only five years, while the Mount Hope North and East was active only seven years, the Lewis Creek field four years, and the Sand River for three years. In some wells, natural gas was produced long after the crude oil had been exhausted.

A major beneficiary of heightened exploration and development in the Denver Basin was the Union Pacific Railroad Company. It had acquired extensive land grants from the federal government for building the Kansas Pacific and the Denver Pacific almost a hundred years earlier. During the Boulder oil excitement of 1901–1902, the Union Pacific, through a local agent, had reminded prospective drillers of wells that the railroad owned the underground minerals for most of the land grant, even when it had sold the surface rights. Because of the narrow confines of the Boulder field, the railroad's claims had not created problems for oil promoters. However, after World War II, some landowners challenged the UP's claim to the underground minerals.

The railroad, commencing in 1892, had reserved to itself subsurface minerals on some sales of surface rights to land. Ten years later, specific mineral reservations were incorporated into all contracts for sales of land. Some surface landowners, hoping to obtain for themselves the proceeds from mineral development, challenged the railroad's right to the oil and gas on grounds that Congress, by terms of land grants, had given railroads rights to "coal and other minerals," which at that time did not include oil and gas. Indeed, some property owners went so far as to seek court rulings that oil and gas were not minerals, which would have exempted them from acquisition by the land-grant railroads. In each instance, the courts ruled for the land-grant railroads. This removed most impediments to the Union Pacific's development of vast mineral holdings in northeast Colorado and elsewhere in the West.

In spite of the rising output of oil, concern for the future prompted major companies to initiate conservation measures in 1955. This effort was designed to prolong production of fields by slowing the loss of underground pressure that was vital for the extraction of oil. The Adena field, of

Morgan County, second largest in the state at that time, was placed under unit operation, or the control of a single operator, in October 1955. Steps were also under way to accomplish the same objective in the Rangely field. Unit operation was a necessary step toward effective introduction of secondary recovery methods designed to increase reserves by as much as 50 percent and to prolong field production. Pure Oil Company, the unit operator for Adena, and Standard Oil of California, in the same role for Rangely, administered highly effective water-flooding operations that slowed, and even occasionally reversed, at least in the short run, declining output. Within about a decade, forty-two of Colorado's fields had active fluid-injection projects. All but three were water-flooded, and the others were combined gas-water injection.

In spite of the successful implementation of conservation programs, Colorado experienced a significant decline in oil output during the 1960s and early 1970s. From 1961 to 1973, the yield shrank by almost 20 percent. In the latter year, twenty fields accounted for more than two-thirds of Colorado's annual production. The deep pool in Rio Blanco County's Rangely field produced 52 percent of the state's total. Wilson Creek, in the same county, was the third-ranking field, with 3 percent of the total. Seven fields in the Denver Basin, led by Peoria (Arapahoe County) and Spindle (Weld County), each with more than 1 million barrels, contributed 13 percent of the state's total.

More than two dozen counties produced crude oil by 1973, but only eleven reported cumulative output in excess of 10 million barrels. Three were in the northwest (Rio Blanco, Moffat, and Jackson) and seven (Washington, Logan, Morgan, Weld, Adams, Larimer, and Arapahoe) were in the northeast. The other one was Fremont County, where the state's oil industry had its origins more than a hundred years earlier. The eleven counties had produced, as of December 31, 1973, more than 1 billion barrels of oil. That represented 97 percent of the state's output of crude oil.

Colorado's refining capacity expanded as new fields added to the output of crude oil. Early refineries were usually small skimming, or topping, plants that removed only kerosene, and later gasoline, from crude oil. They were located in or adjacent to most oil fields of any size. The fuel oil that remained after the extraction of lighter fractions became a glut on the market. Its sale at bargain prices to railroads and other industries undermined traditional markets for coal. Later, rail tank cars and pipelines transported oil from the point of extraction to refineries located in or near urban centers, which were the principal markets for petroleum products. Eventually, the expanding appetite of automobiles for fuel, and the fact that, at best, only about 20 to 30 percent of crude oils could be refined as gasoline encouraged refiners to find ways to increase the yield of readily marketable products.

Refinery residues—oil from which gasoline and kerosene had been removed—could be cracked, to form lighter fractions that could be refined as gasoline. High temperatures, pressures, and catalysts chemically changed the residue into gasoline. This technology was first developed by Dr. William Burton and associates for Standard Oil of Indiana. Burton stills were installed at the United Oil Company's refinery at Florence on the eve of World War I.

In the 1920s, refineries sought new processes that eliminated the Burton system's potentially dangerous combination of high pressure and heat in the distilling process. The most popular of the new methods was developed by Carbon Petroleum Dubbs, with the financial backing of the wealthy Armour family. Dubbs's method of cracking utilized heat and pressure to double the amount of gasoline that could be derived from a barrel of crude.

Other, more sophisticated refining methods improved upon Burton's and Dubbs's processes. Catalytic cracking, developed in the late 1930s and widely adopted in subsequent decades, was the standard for refining until 1960, and remained generally in use after that date. Polymerization, a process of capturing, liquefying, and combining refinery gases, was introduced in the 1930s. Hydrogenation of hydrocarbons gained popularity during World War II because it permitted refiners to increase the output of high-grade gasoline, diesel fuel, lubricants, and a variety of special products for America's armed forces. Alkylation, the application of sulfuric acid to the treatment of crude oil, allowed the industry to meet war-imposed quotas for aviation gasoline. In the 1960s, refineries adopted hydrocracking, which used a catalyst and hydrogen to transform mid-boiling-range oils into high-octane gasoline, jet fuel, and high-grade fuel oil.

The development of new techniques increased the output of more than gasoline, kerosene, fuel oil, lubricants, paraffin, and asphalt. Refinery gases yielded methane, ethylene, propylene, butylene, and the naphthenes from which petroleum chemists synthesized a vast array of useful products. These included antifreeze, medicines, dyes, explosives, nylon and other fibers, plastics, and rubber.

The Rocky Mountain region's first refinery center at Florence, Colorado, processed oil from local wells as early as 1885. All refineries at Flo-

The Continental Oil Company completed construction of a state-of-the art refinery on Brighton Road, north of Denver, in 1931. For many years it processed oil delivered by railcars and pipelines from fields in Wyoming. The Denver Basin emerged as a source of crude oil, as well as natural gas, in the 1950s. Courtesy, Denver Public Library, Western History Department.

rence were affiliated with Continental Oil and absorbed by that enterprise in 1916. The United refinery adopted cracking stills that increased the output of gasoline by reducing heavy fractions of petroleum to the lighter components that could be made into motor fuel. Operations continued at Florence until 1936, by which time the plant had been overshadowed by technically more sophisticated refineries elsewhere in the region.

By 1940, Colorado had sixteen refineries with a total capacity of 145,907 barrels of oil a day. Those outside of Denver were for the most part small skimming plants located in or near oil fields. One exception was the Texas Company's refinery at Craig, which incorporated cracking stills.

Four refineries were in the Denver metropolitan area. The Continental Oil and Bay Petroleum plants, on opposite sides of Brighton Avenue in Commerce City, featured conventional stills and Dubbs cracking units.

Pipelines allowed them to draw crude from, and to send refined products to, distant markets. The Colorado Oil Refining and Colorado Midland Refineries Companies operated smaller refineries. The community's four plants turned out nearly three-quarters of the state's refined products. Before the close of World War II, Oriental Refining and the Skelly Oil Company established refineries in Denver.

The Continental Oil Company (Conoco) built a model refinery north of Denver in 1930. Erected at a cost of $60,000, it was located on a hundred acres of land on Brighton Road, adjacent to the tracks of the Union Pacific, the Burlington, and the Moffat Road. It was later served by pipelines that brought crude oil from Wyoming and transported refined products to distant markets. In 1940, the refinery had a rated capacity of 2,500 barrels daily, and an additional 1,650 barrels could be treated in cracking units. Capacity was increased in 1948. Continental merged its facility with the nearby Empire Petroleum Company's plant in 1965 to acquire additional capacity. By 1970, Conoco's output increased almost 400 percent in response to the growing demand for automotive fuels.

Adjacent to the Conoco site were plants owned by Bay Petroleum Corporation and Oriental Refining Company. Both featured conventional stills and Dubbs units. The Bay plant was owned by Charles Ulrich Bay, who made fortunes in shipping and on Wall Street. His Denver refinery was enlarged during World War II, and again during the Korean War, to a capacity of 12,000 barrels of oil per day. In 1955, Bay Petroleum sold its plant to the Tennessee Gas Transmission Company, which marketed petroleum products under the brand name of Tenneco.

Tenneco had its origins with the formation of the Tennessee Gas Transmission Company in 1943, to build a natural-gas pipeline from the Southwest to West Virginia. After World War II, the company organized the Tennessee Production Company to explore for oil and gas. That small enterprise evolved, in large part through mergers and consolidations, into the Tennessee Gas Transmission Company.

The Oriental Refining Company, with E. M. Stringer, president, produced and distributed petroleum products throughout Colorado. The firm operated thermal plants in Denver and Alamosa with respective treatment capacities of 4,000 and 1,000 barrels a day, and owned oil wells in Colorado and Wyoming. In 1958, Flank Oil Company of Billings, Montana, bought the two plants plus all of Oriental's marketing facilities and crude production for $6.5 million. The new owner continued to operate as the

Oriental Refining Company. Flank Oil was a wholly owned subsidiary of Asamera Oil Corporation of Calgary, Alberta. That firm had been active in Indonesia but shifted its principal petroleum and mining operations to North America in the 1950s.

The Skelly Oil Company built a refinery in the north metropolitan area. It produced a variety of petroleum products, mainly gasoline, for Skelly service stations in Denver and other communities. The plant closed in 1963, at which time the company completed the 204-mile Medicine Bow pipeline to transport petroleum products from Wyoming to Denver. At that time, Skelly transferred its Denver refining activities to El Dorado, Kansas.

In 1973, the only two refineries in Denver belonged to the Continental Oil Company and the Refinery Corporation. They were on opposite sides of Brighton Road in Commerce City. Continental's plant, expanded and upgraded over the years, had been the state's largest for more than four decades. The Refinery Corporation operated what had been the Bay Petroleum Corporation, and then the Tenneco plant. The latter firm sold the facility to General Real Estate and Resource Trust in 1970, and the new owner immediately leased it to Denver oil tycoon John M. King. King forfeited the lease in 1971, unable to secure crude oil with which to operate the plant. That ended his meteoric career and forced the owner to organize a subsidiary, the Refinery Corporation, to process crude oil, most of it from wells in the Denver Basin.

Colorado's third refinery, located at Fruita, was temporarily idle in 1973 because a fire had severely damaged its facilities for making gasoline and fuel oil. The plant, erected by the American Gilsonite Company in 1957, refined gilsonite from the firm's mine at Bonanza, Utah, and transported to the refinery via a seventy-two-mile, 6-inch slurry pipeline. The solid hydrocarbon was a source of gasoline, diesel fuel, asphalt, and high-grade coke. The coke, which was pure carbon, was used in atomic reactors and by the aluminum industry. Sam Gary and Ron Williams, owners of Gary-Western Inc., purchased the Fruita refinery in 1973 and increased the capacity from 8,000 to 10,000 barrels a day. The plant, while idle, was reorganized to process crude oil produced in western Colorado.

That year, Colorado's refineries processed 14.2 million barrels of crude oil, of which 11.1 million came from other states, mainly Wyoming, Montana, and Utah. The principal product was motor fuel for the state's growing number of motorcars. Other products were aviation gasoline and commercial jet fuel, kerosene, distillate fuel oil, residual fuel oil, liquefied gases, lubricants, petroleum coke, asphalt, and road oil. Kerosene was consumed largely for space heating. Distillate fuel oils were diesel and heating oil. Residual fuel oil was industrial heating oil. Liquefied refinery gases included olefins, propylene, butylene, and isobutane, which were used primarily as feedstock by petrochemical plants. Petroleum coke was marketed as fuel and as raw material for manufacturing carbon electrodes for electric furnaces and silicon carbide abrasives.

Colorado's principal producing oil fields in 1973 supplied crude oil to the state's refineries and to those in other states, particularly Utah and New Mexico. Rio Blanco County, with two major fields, accounted for 57 percent of the state's output. Arapahoe and Weld Counties ranked second and third in output, but accounted for only 8 percent of Colorado's production. Rio Blanco led all other counties in cumulative output.

The production of crude oil in the United States had increased dramatically after 1945, and that trend continued for about two decades. At the same time, America's share of world output fell from about 60 percent to approximately 25 percent, and continued to decline in subsequent years. The United States, a net exporter of oil as late as 1947, became, after that date, increasingly dependent on the petroleum resources of other countries. Domestic production failed to keep pace with consumption, forcing refiners to turn to foreign oil. At the same time, the development of the supertanker sharply reduced the cost of transporting oil from the Middle East, and elsewhere, to American ports.

Commencing in the 1950s and continuing into the following decade, energy consumption in the United States increased at a rate of about 3.5 percent a year. After 1965, the pace jumped to 4.5 percent annually. Domestic production of petroleum, as well as other sources of energy, could not keep pace with the growth of demand. Imports of relatively inexpensive foreign oil made up the difference between consumption and domestic production. By 1973, the United States imported about 15 percent of all the energy it consumed. Oil imports alone rose to 24 percent of the annual petroleum supply. When Saudi Arabia reacted to American support for Israel during the Yom Kippur War of 1973 by embargoing some crude-oil exports, the shortage of energy in the United States became critical. The energy crisis marked the end of an era of inexpensive energy and stimulated a national effort to attain energy independence, setting in motion development programs that were to have an enormous impact on

Colorado as well as other western states. Oil exploration and development increased dramatically as high prices for crude petroleum rejuvenated the domestic industry, which had suffered for many years from declining output at home and cheap imports from abroad.

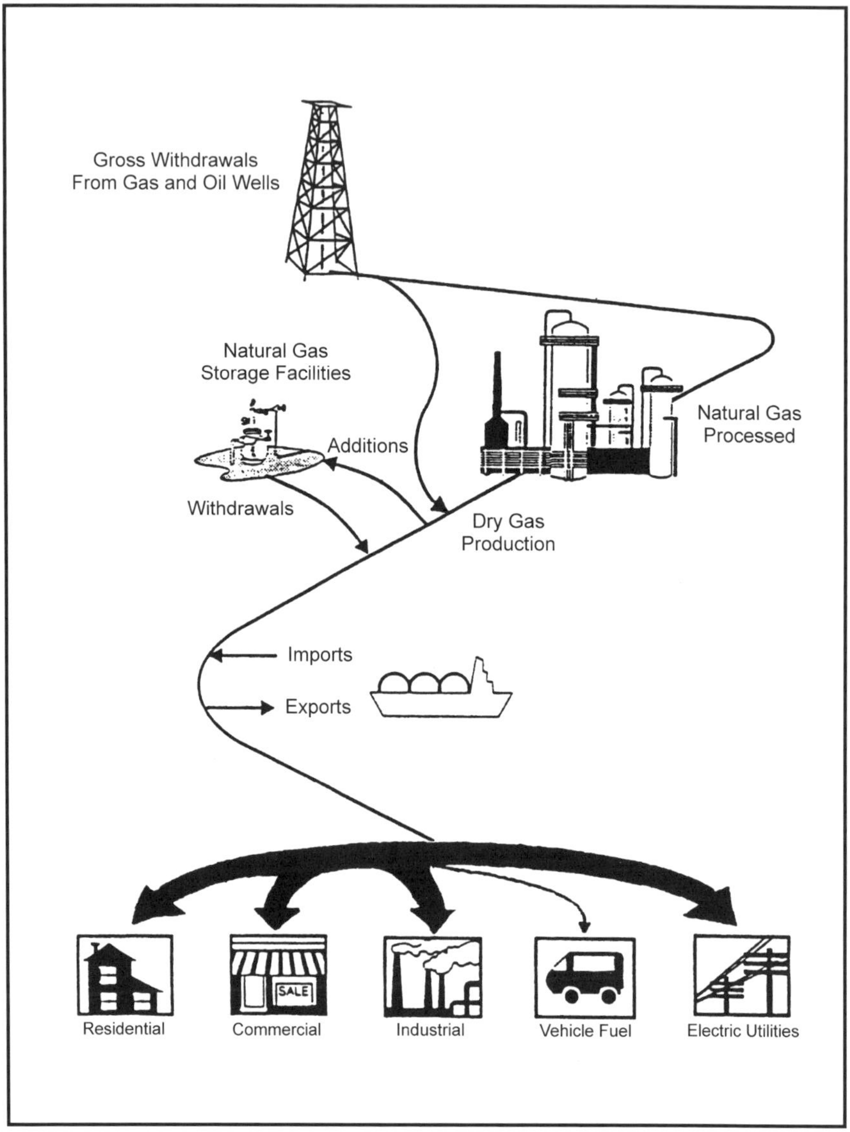

Figure 6. Natural Gas Supply and Disposition in the United States. From U.S. Energy Information Administration, *Natural Gas Annual 1993*.

7

Natural Gas 1930–1973

Colorado's natural-gas industry experienced a roller-coaster ride during the years from 1930 to 1973. Initially, severe economic stagnation discouraged exploration and development of gas fields. In 1941, the last year before the American economy was dramatically distorted by World War II, production remained below the levels of 1930.

There were five active natural-gas fields in Colorado during the Depression decade. The largest, the Hiawatha in northern Moffat County, shipped gas, commencing in 1929, to Salt Lake City, Ogden, and other Utah communities. That year, the Colorado portion of the field accounted for almost four-fifths of the state's output of natural gas. Other producers were the Thornburg field, also in Moffat County, discovered in 1924, but capped for several years because of a lack of markets for the gas; the Garcia and the Berthoud fields, opened, respectively, in 1924 and 1925; and the Craig dome, discovered in 1932.

Gas wells in the Garcia field, in Las Animas County near the Colorado–New Mexico line, about twenty miles southeast of Trinidad, were a source of natural gasoline. The field had been developed for local consumption in the 1890s. In the 1920s, an absorption plant stripped wet gas of gasoline, which was transported by pipeline to Barela station on the Colorado and Southern Railway, then shipped in tank cars to regional markets. The residual dry gas was distributed locally to consumers. The output varied from a high of 96 million cubic feet in 1936–1937 to a low of 54 million cubic feet in 1938. Production ended in 1942.

For a brief time, Las Animas County was a source of helium, a rare inert gas extracted from natural gas. In 1927, the Phillips Petroleum Company completed a gas well in the Red Rocks district, whose output was about 9 percent helium. This did not lead to commercial operations, but it did stimulate exploration in the area. Later the same year, the Girdler Corporation found helium in a gas well on the Model dome near Thatcher. Other wells were put down and a plant erected to extract helium from natural gas. One of two private ventures that produced helium at that time, the company remained in operation until 1937, when the Bureau of Mines opened a large helium plant at Amarillo, Texas. The following year, the federal agency bought all privately owned helium wells and processing plants, making the production of the gas a public monopoly.

At the time, helium was widely utilized in American lighter-than-air craft as a substitute for hydrogen, a highly inflammable gas. The loss of the German airship *Hindenburg* in New Jersey, at the close of a flight from Europe in 1936, pointed to the value of inert helium as the buoyant gas for dirigibles and balloons. Following that tragedy, a policy of no helium sales to other nations gave the United States a de facto monopoly on the construction and operation of airships.

Three miles west of Berthoud, in southern Larimer County, the Midwest Refining Company, a subsidiary of Standard Oil of Indiana, opened a small natural-gas field in 1927. The product was sold to the Public Service Company of Colorado (PSC) for distribution to communities in Colorado and to Cheyenne, Wyoming. During the 1930s, the output was normally about 50 million cubic feet of gas annually.

Gas from the Craig dome, two and a half miles from the town of Craig, the county seat of Moffat County, was piped to that community and distributed for domestic consumption. In 1936, a twenty-two-mile, 3-inch line was laid from the Thornburg field to Craig to augment the declining production of the Craig dome. The following year, another 3-inch line, four miles in length, transported gas from the Iles field to intersect the Thornburg-Craig pipeline. The transmission of gas to Craig was the Thornburg field's first commercial outlet. Sales topped 22.4 million cubic feet in 1936, and climbed to 97 million cubic feet in 1939.

Other gas fields were capable of large commercial production, but lacked markets for their output. For this reason, several in western Colorado remained undeveloped: Bell Rock, Powder Wash, and Thornburg (until 1936) in Moffat County; Piceance Creek, Rangely, and White River in Rio Blanco County; and Garmesa in Garfield County. The outlook did not brighten until after the close of World War II, when new pipelines carried western Colorado gas to large markets on the Pacific Coast.

In Colorado, gas consumption by householders, commercial establishments, and industries grew at a modest pace during the Depression era. Initially, the volume declined from 16.6 million cubic feet in 1931 to 15.8 million cubic feet in 1933, then gradually expanded to 20.8 million cubic feet in 1937. There was a brief contraction during the recession of 1938, followed by continued growth in subsequent years. In 1941, Coloradans consumed 26.6 million cubic feet of gas.

Industries were the principal consumers. Led by operators in oil and gas fields, including manufacturers of carbon black, industries accounted each year for two-thirds to three-quarters of all marketed gas. Householders, whose number increased from 88,140 in 1931 to 102,480 in 1941, consumed about 24 percent, while commercial establishments accounted for about 7 percent of gas consumed annually in the state.

Most of the natural gas marketed in Colorado was imported from Texas, Kansas, and New Mexico. In 1931, total consumption was approximately five times the amount of natural gas produced within the state. In 1941, consumption was about eight times in-state production, indicating a growing reliance on imports.

During World War II, Colorado's production of natural gas surged upward in response to rising demand. The output in 1943 was double what it had been two years earlier, declined in 1944, and dropped to 5,141 million cubic feet in 1945, the last year of the conflict. Only one producing field, Berthoud, was located east of the mountains. Moffat County's Hiawatha, Powder Wash, and Thornburg fields, in that order, were the leaders, and most of their output was shipped via pipeline to Utah. Powder Wash, discovered in 1931 and an oil producer in 1936, remained shut in until 1941, when the Mountain Fuel Supply Company built a pipeline from the field to its Wyoming-Utah transmission line.

After the close of hostilities, the consumption of natural gas increased at a rapid pace. Coloradans used 34.9 million cubic feet of gas in 1945, more than half of which was credited to industry. Households numbering 115,000 consumed 9.3 million cubic feet, and 12,940 commercial enterprises used 3.1 million cubic feet. Consumption increased to 88.7 million cubic feet in 1950, when 170,000 households used 22.2 million cubic feet and 23,000 commercial establishments consumed 11.1 million cubic feet. As in the past, industries were the largest consumers, taking 55.4 million cubic feet.

There was a proliferation of new producers in the 1950s. At the same time, longtime shut-in fields began shipments to markets. Much of the growth in output was west of the Continental Divide, particularly in La Plata County, where the Barker and Ignacio fields became the leaders in the production of dry gas. Most of this was exported via new pipeline systems to the Pacific Coast. Equally important, the Eastern Slope of the state, long dependent on Texas and Kansas for natural gas, became a major producer in its own right. The Denver Basin was explored and developed as a source of wet gas, which was processed to remove natural gasoline and other liquid petroleum products before the residue was sold to the Public Service Company for distribution to urban residents. The dramatic expansion of natural gas, as well as crude petroleum, production forced the state to assume regulatory activities designed to ensure that producers operated in the best interest of consumers.

In the nineteenth century, when gas was distilled from coal, municipalities had exercised a measure of control over the manufacture and distribution of illuminating fuel by granting franchises that protected their holders from competition in local markets, and authorized the right of eminent domain in order to acquire surface or underground rights to property on which to construct distribution systems and related facilities. In return, local governments were assured ample supplies of gas to meet the needs of the community at rates fixed for specified periods of years. Inevitably, utilities became key players in municipal politics, ruling out the likelihood that local governments could effectively protect citizens against unfair rates and arbitrary practices by franchised electric, gas, water, and transportation monopolies.

The inability of municipalities to control the activities of utilities in the public's interest led to demands for the state to exercise its regulatory authority. The high cost of building and maintaining utilities forced companies to seek charters, or certificates of incorporation, as a way of mobilizing large amounts of capital while limiting risks. This could be done only with the approval of state governments, which by law not only issued charters but also exercised responsibility for seeing that corporations adhered to the terms of those documents. During the progressive era of the early twentieth century, a time when large enterprises were accused of a variety of abuses, many states created utilities commissions to protect the public's interest.

The Colorado General Assembly was reluctant to move in this direction. In 1908, the legislators ignored a request from Governor John Franklin Shafroth to create a state agency to regulate railroads. Two years later, responding to mounting public pressure, it created a commission that lacked real authority to regulate the railroads. In 1913, the growing influence of progressivism in the Colorado House produced a law that authorized a Public Utilities Commission (PUC) to regulate rates and services of utilities—transportation, electric, gas, telephone, and water—that operated in intrastate commerce. Initially, the agency's focus was largely on transportation enterprises. By the 1920s, it closely monitored electric, gas, and telephone corporations.

The Colorado PUC was primarily concerned that prices charged for electricity and manufactured (later, natural) gas reflected the quality and quantity of services rendered by utilities. The agency insisted that services had to be reliable and available without discrimination to all customers at reasonable rates. The commission's attention largely centered, in fact, on the regulation of rates.

Franchised monopolies were permitted to set rates that ensured the recovery of actual costs, plus just and reasonable returns on investments. However, the determination of costs, and their application to rates for services, invariably proved to be a very difficult task. The public frequently complained that rates were too high, while the utilities insisted that they were too low. Alleged abuses by producers of manufactured gas, who were also the local distributors, were common complaints. Mergers and consolidations within the industry reduced competition and, according to critics who professed to speak for consumers, enabled gas retailers to charge whatever the markets would bear.

At the urging of Congress, the Federal Trade Commission launched a comprehensive investigation of natural-gas production and distribution. A massive report released in 1935 revealed that the industry was controlled by a few enterprises, that enormous waste occurred in producing fields, and that charges for gas, particularly those imposed by transmission companies, were often excessive. Congress, influenced by the report, approved two major reforms—one in 1935, the other three years later. The first, the

Public Utilities Holding Company Act, sought to restrain the growth of utility monopolies. It empowered the Federal Power Commission (FPC), which had been created in 1920 to administer the Federal Power Act, to regulate rates and business practices of interstate electric utilities. Structurally, holding companies were limited to two levels of enterprises; an amendment in 1938 reduced operations to a single integrated system confined to the sale of power.

A second reform asserted the federal government's right to regulate the interstate gas industry. By terms of the Natural Gas Act, the Federal Power Commission was authorized to take over some of the regulatory functions previously exercised by state and municipal agencies. The FPC was given jurisdiction over rates, accounting practices, and the construction or abandonment of pipelines and related facilities. The act applied only to the transportation of gas in interstate commerce and its sale for public consumption. The commission was denied authority over field production and intrastate distribution of gas. It could, however, control both imports and exports.

Federal regulatory functions increased from 1941 through 1945 because of the critical role of natural gas as a source of energy for war-related industries. The Office of Petroleum Coordinator for War Production sought to ensure ample fuel for production of steel, armaments, munitions, and other implements of war. The Office of Price Administration and the Office of War Utilities, divisions of the War Production Board, manipulated gas production and distribution, while at the same time holding the line on prices. The FPC gained from Congress additional authority over rate-making and the construction and operation of interstate pipelines.

In 1945, the U.S. Supreme Court supported a larger regulatory role for states. The high tribunal affirmed the authority of the Texas Railroad Commission to regulate output and to set prices at the wellhead. This encouraged states to create agencies, usually called "oil and gas conservation commissions," to impose rules and regulations designed to prevent waste of crude petroleum and natural gas.

Typical of this development in the mountain region was the Colorado Oil and Gas Conservation Commission. Created by statute in 1951, at a time when the rapid expansion of oil and gas production raised concerns about inefficiency and waste, the agency was appointed by the governor, with the advice and consent of the state senate. The five commissioners were empowered to promulgate rules and regulations to prevent waste in the production and distribution of gas and crude oil. Under the supervision of a director, staff members could conduct investigations and inspections to ensure compliance with established policies.

Colorado's gas industry, whose growth during the war years was sharply limited by the inability to open new wells, expanded at a rapid pace following the close of the conflict. Rising demand in the states of the Pacific Coast, Midwest, and East accelerated the pace of exploration and the construction of long-distance transmission lines. With markets ensured, gas production soared throughout the state.

The rising demand for natural gas sustained an accelerated schedule of drilling activities throughout western Colorado. The first important interstate outlet was a 1,200-mile pipeline completed in 1947 by the El Paso Natural Gas Company from the Permian Basin, in Texas and New Mexico, to California. The relatively isolated San Juan Basin, which extended from northwestern New Mexico into Colorado, Utah, and Arizona, experienced intense exploration and development as soon as the pipeline network was in place to carry natural gas throughout much of the Southwest, and to intersect with El Paso's trunk line to the Pacific Coast.

Production soared elsewhere in western Colorado. Piceance Creek and other shut-in fields began commercial sales in 1953, when the Western Slope Pipeline Company completed a gas line to Grand Junction. The Pacific Northwest Pipeline, from the San Juan Basin to Seattle, provided additional outlets for gas fields in Colorado, eastern Utah, and southwestern Wyoming. A merger in 1959 temporarily transformed the Pacific Northwest line into El Paso's northern operating division. Five years later the merger was ruled invalid by the Supreme Court.

Interstate transmission lines stimulated exploration and discovery in southwestern Colorado, where Barker Creek and Ignacio became large producers. Barker Creek was an extension of the New Mexico field of the same name. Stanolind Oil and Gas Company's (Standard Oil of Indiana) discovery well, Ute Indian No. 1, completed in 1950, had the potential of producing 15 million cubic feet of gas per day. Subsequent development made Ignacio-Blanco the state's largest gas field.

Expanding pipeline systems opened additional markets in the 1960s. Denver-based Rocky Mountain National Gas Company tapped into producing fields to supply consumers in Glenwood Springs, Basalt, Carbondale, and Aspen. In 1968 the firm extended its pipeline system to Delta and Montrose. That year, Cascade Natural Gas Corporation built a large plant to process gas from the Dragon Trail field, with the residue assigned to the Mountain Fuel Supply Company's transmission line to Utah. Continental Oil put down additional wells in the Dragon Trail field to supply Cascade's need for 67 million cubic feet of gas a day. In 1969, Chadbourne Corporation, a Texas firm, began operation of a refrigeration-absorption plant that extracted large amounts of propane and natural gasoline, with the residue assigned via the transmission system to the Pacific Northwest.

Natural gas was either dry or wet. The former contained no liquid petroleum products and, unless other impurities were present, could be marketed without processing; the latter had to be processed to remove liquid components. A large part of the gas produced in the Colorado portion of the San Juan Basin was dry. Other important sources of dry gas were portions of the Piceance Creek, Hiawatha, and Powder Wash fields. The Continental Oil Company and Western Slope Gas Company opened a processing plant north of Grand Junction in the 1960s to handle gas from Douglas Creek, Dragon Trail, Garmesa, and other fields. The state's largest oil field, Rangely, produced, as well, a large volume of wet gas. After natural gasoline, butane, and propane were removed, most of the residue was injected to maintain the pressure in the underground oil reservoir.

While western Colorado enjoyed an abundance of natural gas, Denver and other communities along the Eastern Slope relied on pipelines built in the 1920s and 1930s to transport gas from the Texas Panhandle and Kansas. By the postwar years, these were not sufficient to keep pace with urban industrial growth. In an effort to meet anticipated demand, the Colorado Interstate Gas Company constructed a 240-mile, 20-inch transmission line from the Hugoton field in southwestern Kansas to the Mile High City in 1947. This was distributed by the Public Service Company of Colorado to customers in northeastern Colorado and Wyoming.

In the 1950s, locally produced natural gas augmented imports as the area north and east of Denver became the focus of exploration and development. Wildcatters tapped the oil and gas potential of the vast area in 1950. A joint venture by the Plains Exploration Company and British-American Oil Producing Company successfully drilled discovery wells in the Armstrong and Yenter fields of Logan County. At about the same time, Adams Drilling and the J. M. Huber Corporation discovered the Lee field in Morgan County, Shell Oil opened the Mount Hope field in Logan County, and W. E. Atkinson and Associates found gas in the Loveland field in Larimer County. The Yenter field was linked to Denver by a 12-inch transmission line in 1951. Three years later, Continental Oil, Pure Oil, and other companies completed gathering systems and gas-processing plants for the Adena and Fort Morgan fields in Morgan County, the Badger Creek and Little Beaver in Adams County, and the Bobcat in Washington County. After natural-gas liquids had been removed, the residue was sold to Natural Gas Producers Inc., a subsidiary of Colorado Interstate Gas Company, for distribution to utilities. Some gas was sold to the El Paso Gas Company for transmission elsewhere in the West.

The pace of exploration for oil and gas in the Denver Basin accelerated in the summer of 1969, when the Amoco Production Company leased about 2.8 million acres from the Union Pacific Railroad Company. Amoco conducted tests on some of the land, and entered into contracts for the same purpose with independents. Under the latter arrangement, Thomas G. Vessels, a Denver operator, completed No. 1 Grenemyer in 1970, the discovery well in the Wattenberg field of southern Weld County. The well produced 484 barrels of oil and 26,630,000 cubic feet of natural gas daily. Panhandle Eastern Pipe Line Company built a gathering system to collect gas from wells in Wattenberg and other fields for processing at a compressor station twenty miles south of Greeley. Within two years the field had expanded into Adams County, and it was the state's largest producer of wet gas. Eventually it extended into Boulder County. After liquid petroleum components had been removed, the residue was sold to the Public Service Company of Colorado for distribution to consumers in the Denver area.

Vessels and Amoco discovered, in addition, the Peoria field, in Arapahoe County forty miles east of Denver. The exploratory No. 1 Price produced 660 barrels of oil and 20,000,000 cubic feet of wet gas a day, marking the opening of another large gas field in eastern Colorado. That same year, Vessels's No. 1 Price-Brother opened the Peoria field as a major source of oil. The discovery well produced 4,702 barrels of oil and 7,100,000 cubic feet of wet gas per day. Production was restricted until Amoco's gas-processing plant began operation in March 1971. The output immediately soared from 4,400 to 9,353 barrels of oil and to 2,764,000 cubic feet of gas daily, ranking the Peoria field, at that time, second in the state in the production of crude oil and natural gas. Rangely led all rivals in both categories. The processed gas was sold to Colorado Interstate Gas Company.

Coincidental with the growth of both production and demand, the Colorado industry developed underground facilities to store natural gas.

This practice allowed transmission companies and utilities to cope effectively with expanded demand for gas caused by extreme cold weather. When the temperature dropped to zero and below in the winter months, customers used a volume of gas that exceeded the capacity of transmission lines. At these times, industries on interruptible service had to shut down or turn to alternative fuels. In order to avoid interruptions to households and commercial establishments, transmission companies and utilities created storage facilities from which they could draw gas to supplement the volume normally available for distribution.

Five underground storage facilities were created in the 1960s. The largest was Colorado Interstate Gas Company's Fort Morgan field, where exhausted oil and gas wells served as temporary storage for up to 13.9 billion cubic feet of gas. The Kansas-Nebraska Natural Gas Company used the Springdale field in Logan County in a similar fashion. The Public Service Company of Colorado converted the abandoned Leyden coal mine in Jefferson County, west of Denver, to storage. The Western Slope Gas Company, a subsidiary of Public Service Company of Colorado, stored gas in the once-active Asbury Creek gas field of Mesa County, and in the Home Creek field of Montezuma County.

The Leyden facility was unique in Colorado. The Public Service Company converted the abandoned coal mine, seven miles northwest of Arvada, into a holding area for gas. Tests began in 1959 to determine the site's storage potential. After gaining the approval of the Colorado Oil and Gas Conservation Commission, the company, commencing in 1960, stored 1–3 billion cubic feet of gas during the summer months that was available for the metropolitan area during periods of peak demand. In December 1961 the Leyden mine proved its worth, providing 20 to 50 million cubic feet of gas a day during a record-setting cold wave.

After forty years of operation, the Leyden storage facility became the target of complaints from residents who lived nearby in west Arvada. Leakage of gas raised safety issues, prompting demands from neighbors and city and county officials that the Public Service Company end gas storage in the mine. Faced with lawsuits and threats of more, the PSC initiated steps to shut down the plant no later than 2003.

The Western Slope Gas Company made Asbury Creek, a depleted gas field near Grand Junction, a storage facility to do more than respond to peak-load demands. It stored high-Btu natural gas from the Dragon Trail and Lower Horse Draw fields, from which liquid petroleum products had been extracted by the Continental Oil Company's plant at Fruita. The gas was used to stabilize the Btu content of fuel delivered to customers by blending it with low-Btu gas from the Garmesa field.

In 1970, the United States derived approximately 30 percent of its energy from natural gas. The fuel's post–World War II growth in popularity had resulted from its attractiveness in terms of cost, convenience, and availability. Its high heat content made it relatively inexpensive, as well as easy and clean to use. Unlike coal and oil, gas was largely nonpolluting and, in environmental terms, it was the ideal fuel for consumers in urban centers.

Increased use in the United States raised questions about future availability. According to some estimates, the demand for gas in 1975 would exceed supply, at then-current rates of growth, by 5 trillion cubic feet, or about 20 percent. This focused attention on expanding output by means of a complex strategy, one part of which called for importing liquified natural gas (LNG) from other parts of the world. At the same time, the nation's vast coal deposits were to be utilized as a source of manufactured gas to supplement the supply of the natural variety. In order to increase production from known fields, huge volumes of gas were to be obtained by fracturing the sandstone formations in which they were trapped. Finally, the deregulation of gas exploration, production, and distribution was, according to free-market advocates, the one sure way to increase supply faster than the growth of demand.

As in the case of crude petroleum, supplies of natural gas could be augmented by imports. Obvious foreign sources were Algeria and Russia. Gas could be liquefied at a temperature of minus 259 degrees Fahrenheit and transported in super-insulated tankers equipped with compartments that resembled huge thermos bottles. Stored in the United States until needed, 1 cubic foot of compressed liquefied gas expanded at normal temperatures to about 600 cubic feet.

In 1969, the importation of liquefied natural gas from Algeria was sponsored by El Paso Natural Gas Company, which, for this purpose, entered into partnerships with other gas transmission companies. The marketing of imported gas, according to the industry's critics, was a way of circumventing federal price controls on gas produced in the United States and marketed in interstate commerce. Because of innumerable delays, initial deliveries did not occur until 1976.

Imports of liquid natural gas from outside the continental limits of the United States largely ended in 1978, by which time the supply of domestic natural gas in the newly deregulated domestic market proved ample. This was a severe financial blow to El Paso and its transmission associates. Ultimately, this liquefied natural-gas fiasco contributed to El Paso's acquisition by Burlington Northern Inc. In spite of the many problems associated with imports, LNG was used on a small scale on the Atlantic seaboard. Imports from Alaska and Indonesia served some parts of the Pacific Coast as long as natural gas was in short supply.

Potentially more promising as a way of augmenting natural-gas supplies was the synthesis of coal to produce a high-quality substitute. This required far more than reerecting and operating the plants that once manufactured gas for America's municipalities. That product yielded, at most, gas with heat values of about 300 Btu per cubic foot. Synthetic gas, to be cost-effective, had to yield approximately 1,000 Btu per cubic foot. The expense of making the latter was significantly higher, and the requisite technology was far more sophisticated than the gashouse retorts of an earlier day.

In an effort to encourage development of the requisite technology, the Office of Coal Research (OCR), within the United States Bureau of Mines, funded numerous investigations by private enterprises and research institutions commencing in the 1960s. An early project was undertaken by the Consolidation Coal Company, then a Continental Oil Company subsidiary, in West Virginia and later in South Dakota. That effort indicated that approximately 250 million cubic feet of gas could be produced daily from High Plains lignite at a competitive cost of less than 50 cents per million Btu. A pilot plant was constructed near Rapid City to determine the feasibility of the gasification project.

Another OCR contract was awarded in 1968 to the University of Wyoming's Natural Resources Research Institute for a five-year study of methods for converting coal to gasoline and other petroleum products. The institute found that a coal gasification pilot/demonstration plant would cost about $15 million, and a commercial facility about twice that amount. In pursuit of the former, OCR underwrote five programs. While suggestive of new directions for future applications, the projects were not capable of preventing an anticipated gas shortage. At best, the pilot plants had combined capacities of about 90 billion cubic feet per year, with attendant high costs and risks of air pollution. They could produce only about 5 percent of the gas consumed in America. In light of an expected shortage of 5 trillion cubic feet by the mid-1970s, the gasification of coal, given the technology of that time, was not a viable option for narrowing the gap between demand and supply.

Another possibility for enhancing output was nuclear stimulation. The industry had used explosives and hydraulic fracturing to stimulate production of gas wells by shattering underground rock formations that held what operators called "tight gas." The results had been unimpressive because of the small area that could be treated with the available technology. However, engineers suggested that an underground nuclear explosion could ensure massive fracturing and correspondingly large gas yields.

The Atomic Energy Commission (AEC, later the Nuclear Regulatory Agency), created in 1957 a Division of Peaceful Nuclear Explosions to pursue nonmilitary applications for atomic energy. The project was called Operation Plowshare. The agency considered and eventually rejected numerous proposals, including one for nuclear explosions to enlarge the Panama Canal, or to excavate a new Central American interocean waterway. In pursuit of a more feasible goal, the AEC authorized the use of nuclear power to stimulate the production of natural gas and oil from underground rock formations.

Massive underground stimulation seemed particularly applicable to the Rocky Mountain states, where an estimated 600 trillion cubic feet of gas was locked in sandstone and shale. These formations had to be shattered in order to obtain a high rate of recovery. In association with private enterprises, the AEC carried out three underground explosions. The first two projects—Gasbuggy in 1967, fifty miles east of Farmington, New Mexico, and Rulison, in Colorado's Piceance Basin in 1969—yielded indifferent results. A third, also in the Piceance Basin, featured multiple explosions in 1973.

The first two projects expanded output, but not enough to justify the great expense and risks peculiar to subsurface nuclear explosions. At the New Mexico site, a 26-kiloton nuclear device was detonated underground adjacent to gas-bearing formations where massive fracturing could significantly increase production. Prior to the test, a single well produced at the rate of 10,000 cubic feet of gas per day. Afterward, the yield increased to 100,000 cubic feet per day, but the gas was heavily laced with radioactive particles and unsuitable for commercial development.

Project Rulison, in the Grand Valley of Garfield County, employed the technology used for Gasbuggy, with the addition of a boron carbide shield to reduce radiation from the detonation. A 40-kiloton nuclear device was exploded on September 10, 1969, at a depth of more than 8,000 feet, with the objective of releasing trillions of cubic feet of gas trapped in tight sand in the Rulison field. Prior to the blast, employees were ordered out of underground coal mines within a radius of forty miles of the test site. Project Rulison, as in the case of Gasbuggy, was a technical success, but the high level of tritium and krypton, the principal radioactive contaminants released by detonation, made the natural gas of questionable value for domestic or industrial consumption. According to the natural-gas industry, tests of the Rulison gas revealed low-level radiation that posed no risk to public health and safety, but governmental agencies reached the opposite conclusion and ordered a permanent seal on the well in 1976.

The explosive devices for a third nuclear-stimulation test were designed to reduce the yield of radiation particles. Sponsored by twenty-five oil companies, the Atomic Energy Commission, and the Bureau of Mines, the test was scheduled to take place in 1972. CER Geonuclear Corporation completed a feasibility study, while the Bureau of Mines made test borings in Colorado and Wyoming in search of a suitable test site. One was finally located in the Piceance Creek Basin in Rio Blanco County. However, political opposition delayed and then temporarily blocked the experiment.

Rio Blanco was a disappointment, in part because the three shots were not properly placed to create a single chimney of shattered rock. In addition, engineers miscalculated the amount of gas available in that portion of the Piceance Basin. All three sites were sealed. The failure of Rio Blanco prompted the cancellation of another test, Project Wagon Wheel, in Wyoming. By that time, widespread public opposition to nuclear projects had developed. In 1974, a statewide referendum in Colorado banned nuclear-stimulation projects within the state's boundaries, unless approved by a referendum of the electorate.

The widening gap between supply and demand prompted, by 1970, the natural gas industry to renew its request that Congress end federal regulation of the interstate market. At issue was the Natural Gas Act of 1938, which gave to the Federal Power Commission supervision over gas-transmission companies that supplied interstate markets.

Originally, the FPC regulated only prices for gas delivered by interstate pipelines to utilities for local distribution. It did not attempt to fix prices for gas sold at the wellhead, unless the producer and the transmission pipeline company were the same enterprise. However, in 1954 the U.S. Supreme Court, in a case brought by the State of Wisconsin against the Phillips Petroleum Corporation, ruled that the Natural Gas Act of 1938 required that if the gas was sold to pipelines for interstate commerce, the FPC could regulate prices at the wellhead.

In carrying out the mandate of the Court, the FPC, because of the difficulty of determining costs, set tentative prices that remained in effect for the next decade. This had the effect of imposing a freeze at rates that discouraged exploration and the expansion of production. This, according to industry spokesmen, was primarily responsible for the failure of supply to grow apace with demand. The remedy, they insisted, was simple. End the regulation of wellhead gas destined for interstate commerce, and incentives, in the form of higher prices, would lead to increased output.

In support of this argument, they pointed to the impact of a dual price structure on gas in intrastate commerce. In Texas, a major producing state, there was no shortage of gas because the wellhead price was substantially above that for gas sold in interstate commerce. Therefore, Texas gas was, whenever possible, consumed only at home rather than shipped to other states.

What the industry wanted was an end to federal regulations. What it gained in the short term was a temporary price increase. Eventually, Congress phased out controls over a period of eight years. It remained to be seen if, as industry spokesmen claimed, the end of regulation would lead to increased exploration and production in Colorado and elsewhere in the nation.

In Colorado, as in the rest of the country, the output of natural gas had increased steadily in the 1950s and continued to climb to a peak of 136.7 billion cubic feet in 1966. Sufficient new discoveries did not occur in the 1960s to compensate for the depletion of existing fields, with the result that output declined gradually to 105.8 billion cubic feet in 1970. An anticipated continued drop in output did not happen because of the opening, at that time, of new fields in the Denver Basin. The largest was Wattenberg, followed by Peoria, Third Creek, Sprindle, and Lowry. Within three years, the state's total production had rebounded to 137.7 billion cubic feet, and it continued to rise, with some minor fluctuations, for the next quarter of a century.

In spite of the surge in output, the production of natural gas did not keep pace with advancing demand, and Colorado remained, as it had since the 1920s, dependent on supplies from other states. In 1970, for example, the state's production of natural gas equaled only 37.5 percent of consumption. There was noticeable improvement by the mid-1970s, when production in-state equaled 52.6 percent of consumption.

Consumption increased from 224 to 282 billion cubic feet in the years from 1965 to 1970, peaked at 308 billion cubic feet in 1975, and declined during the rest of the decade. Approximately one-third of consumption was by residential customers in the early 1970s. By 1975, commercial and industrial customers each accounted for about one-fourth of the gas consumed in the state. Industrial use, including electric utilities, fell off as federal programs commencing in 1973 required the substitution of coal for natural gas.

Nationwide, producers and consumers agreed that something had to be done to increase output, but there was no agreement on what action should be adopted. That impasse was overcome in October 1973, when an Arab oil embargo prompted a national commitment to pursue energy self-sufficiency. One important aspect of that goal was the expansion of production, in part from underground reservoirs, and also from the distillation of coal to produce a pipeline-quality product that approximated, in Btu, the heat content of natural gas.

8

Synthetic Fuels 1917–1973

Throughout most of the twentieth century, the United States had an abundance of fossil fuels in the form of coal, petroleum, and natural gas. Coal was a seemingly inexhaustible resource, but occasionally unanticipated shortages of petroleum and natural gas prompted calls for the production of synthetic fuels derived from coal and oil shale. Some of the technology to produce synthetic fuels had been developed long before there was a commercial petroleum industry. Kerosene, for example, was known as coal oil because it had been made at one time from the solid fossil fuel. Furthermore, there had been a thriving market for gas distilled from coal in the nineteenth century, and its widespread usage continued in some parts of the country, and abroad, until after World War II.

From time to time in the twentieth century, fears of petroleum shortages caused government bureaucrats and petroleum industry officials to ask Congress to create research programs designed to find cost-effective techniques for extracting oil from shale. On most occasions, Congress responded with legislation that authorized experimental programs designed to pave the way for the commercial production of synthetic fuels. However, the scarcity of petroleum always proved to be of brief duration, giving way to a condition of abundant cheap supplies that undermined and eventually forced the abandonment of synthetic-oil programs.

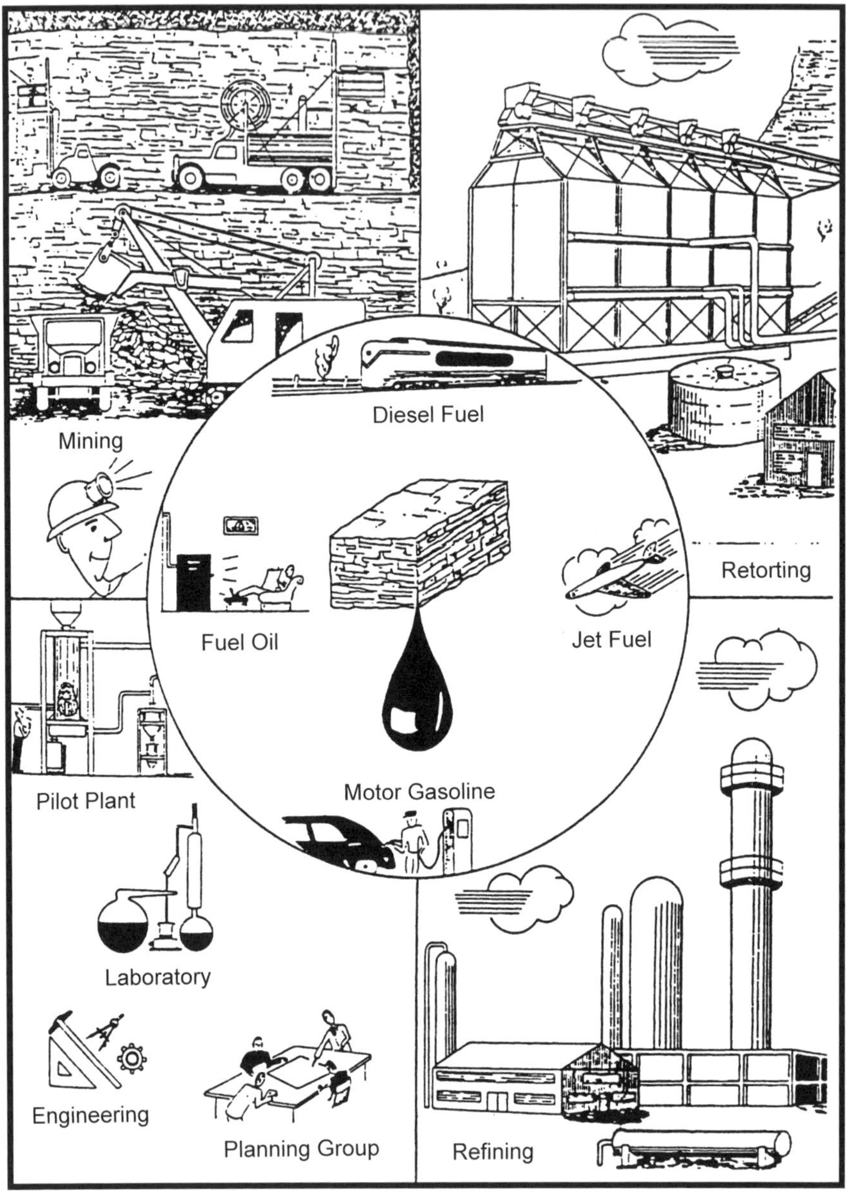

Figure 7. Production and Disposition of Oil From Shale. From U.S. Bureau of Mines, *Report of Investigations 4771* (1951).

Oil shale is the term used to identify organic marlstone that contains kerogen, a flammable substance that is similar in appearance and odor to petroleum. It seems likely that the two had common origins. However, in the distant past, kerogen-bearing shale was not subjected to the extreme heat and pressure that turned organic matter into petroleum. Kerogen vaporizes when shale is heated to temperatures of more than 700 degrees Fahrenheit. The condensed vapor is a black viscous liquid that can be refined to make liquid fuels and lubricants. Additionally, it is an excellent substitute for petroleum for the manufacture of numerous medicines and plastics.

Shale has been a source of oil for almost two centuries. The commercial extraction of oil from shale began in France in 1838, but by far the most advanced industry of that time developed in Scotland, where at one time more than fifty companies processed shales. They produced oil that was distributed throughout the British Isles, plus large amounts of ammonia sulfate, which was marketed as fertilizer. Similar industries were active in the United States prior to 1860, and in the twentieth century flourished in Europe, Latin America, and other parts of the world. Production ended in most countries after World War II when abundant, cheap crude petroleum precluded the need for an expensive alternative.

Oil shales of commercial potential are located in thirty states in the United States, but by far the most important deposits are in the intermountain West. The Green River formation of northeastern Utah, southwestern Wyoming, and northwestern Colorado covers an area of approximately 17,000 square miles and contains an estimated 2 trillion barrels of oil. The richest sections of these deposits are in Colorado, which contains a minimum of 400 billion tons, with an average yield of 25 gallons of oil per ton of shale. In addition, there are an estimated 120 billion barrels of oil in the shales of Utah, and another 12 billion barrels in Wyoming.

The area that comprises the Green River formation was a salty inland lake when the shale beds were formed approximately 36 million years ago. Today the shale is in horizontal and angular strata that range in thickness from 20 inches to more than 100 feet. Some strata are exposed, while others are hidden at various depths beneath the surface. Highly visible are massive escarpments lying north of the Colorado River between the towns of Rifle and De Beque, and along tributary streams that cut through the shale formations. Perhaps the most interesting is the Mahogany Ledge, located along Parachute Creek, a kerogen-rich zone 100 feet in thickness.

Old-timers in western Colorado tell several stories about how local residents became aware of the peculiar qualities of what they called the "rock that burns." A popular version relates to a schoolhouse that was constructed with a fireplace fashioned from rock that was quarried on the site. A ceremony marking the completion of the structure required use of the fireplace. The next morning the structure was a mound of ashes, and an investigation revealed that the fire had ignited the rock of the fireplace, and the flames devoured the wooden structure.

A similar tale relates that a settler erected a cabin near Collbran. He built a large structure, complete with a massive fireplace fashioned from slabs of stone collected on the slopes of a nearby mountain. During the first snowstorm, the pioneer built a roaring fire of pine logs to warm his home. Much to his dismay, the stone caught fire, emitting the odor of kerosene. He subsequently learned that he had located his residence next to a mountain of oil shale.

The U.S. Geological Survey (USGS) initiated the first comprehensive study of western shale and its potential as a source of oil. Geologists David T. Day and Elmer G. Woodruff examined deposits throughout the Uinta Basin and published a report, *Oil Shale of Northwestern Colorado and Northeastern Utah*, in 1914. Two years later, another USGS geologist, Dean E. Winchester, published *Oil Shales of Northwestern Colorado and Adjacent Areas*. The appearance of the second report coincided with and contributed to the start of the first effort at large-scale extraction of oil from shale in the intermountain West. Winchester estimated the area contained 300 million barrels of oil. A year later, he raised the estimate to 400 million barrels, at a time when it appeared that the nation's supplies of petroleum were incapable of keeping up with growing demand for fuel for internal-combustion engines.

Reports of a pending petroleum shortage triggered a rush by prospectors and fortune seekers to claim oil-shale lands in the intermountain West. The first placer claims, staked under the provisions of the Mining Act of 1872, had been filed in Colorado's Piceance Creek Basin in 1910. In the next decade, approximately 40,000 claims were filed on several million acres of shale land. The peak years of activity were 1916 to 1918. The U.S. General Land Office, in the Department of the Interior, issued patents for only a small percentage of the claims, but that did not slow the region's first oil-shale boom.

There have been three attempts at large-scale development of the oil-shale deposits of the Green River formation. An on-again, off-again oil-from-shale industry first emerged in 1916 or 1917, and continued throughout most of the decade of the 1920s. There was a revival of interest in the utilization of oil shale during World War II and for about ten years after the close of that conflict. A final period of intensive development coincided with the energy crisis of the 1970s and 1980s. Each period of activity had its origin in the perception that supplies were diminishing at a time when demand for petroleum products was increasing. Each oil-shale boom ended because of the expansion of available petroleum supplies. In each instance, there was an almost total collapse of the fledgling oil-shale industry.

The initial boom coincided with the abnormal demand for oil induced by World War I, a time when America's consumption of petroleum threatened to outstrip available supplies. As the price of crude oil jumped from less than 50 cents to almost $2 a barrel during the years 1915 to 1917, industry analysts looked for low-cost alternatives. Oil shales, particularly those in western states, were, because of reports recently published by the USGS, recognized as potential sources of oil. Despite uncertainties about the technology and extraction costs, some promoters confidently believed that oil from shale could be marketed in competition with crude petroleum.

The Piceance Creek Basin of Garfield and Rio Blanco Counties, about two hundred miles west of Denver, became the center of oil-shale development. The towns of Grand Valley (later renamed Parachute) and De Beque served as local headquarters for many prospectors who, driven by dreams of immediate riches, rushed to stake claims throughout the area. This stimulated business activities in Glenwood Springs and Rifle, to the east of the shale-rich basin, and in Grand Junction to the west, the area's principal commercial centers. Claims were also filed in great numbers on deposits in the Utah and Wyoming portions of the Green River formation. Some of the claims were sold and resold many times as speculators enriched themselves at the expense of gullible investors.

Entrepreneurs, promoters, and gamblers organized numerous commercial enterprises to extract oil from shale. Their goal, if advertisements in newspapers and investment journals are accepted at face value, was to build and apply modern technology to the tasks of mining shale, heating it in retorts of various designs to achieve a high yield of kerogen, which could then be refined to make petroleum products. Some companies boasted that they possessed the requisite technology; all they needed was money. Investors were assured large returns. In fact, many, if not most, of the new firms did nothing more than sell shares of stock to eager buyers who anticipated fortunes from oil shale. Perhaps two dozen companies made honest efforts to acquire and apply advanced technology to the extraction of oil from shale at costs that would permit their products to compete successfully with petroleum in the fuel market.

Approximately 250 companies were organized to process the shales. Of these, about 150 chose to do business in Colorado; however, only a few possessed the capital and technical knowledge to build plants, mine shale, and produce oil. The remainder were little more than paper enterprises whose organizers hoped to profit from the sale of stock. The Colorado Bureau of Mines identified 64 out of the 150 companies as "operating" enterprises. All 64 were in business during part or all of the years from 1920 to 1927. While the term "operating" was not defined, analysis of the agency's reports suggests that it meant enterprises that actually intended to engage in mining or to process shale. All of these endeavors were located in Garfield and Mesa Counties, in and around De Beque, Grand Valley (Parachute), Rifle, and Collbran.

Many of the companies may have built facilities for mining or retorting, but only a few actually extracted oil from shale. The experience of the American Shale Refining Company typified the difficulties of doing business during the oil-shale boom. Organized in 1917, the enterprise filed claims on 4,500 acres north of De Beque. Plans called for the extraction of more than 200 barrels of oil daily, and profits were projected at about $500,000 a year. The leading stockholders had been active in Wyoming's booming petroleum industry. The firm appeared to have the capital, technology, and managerial skills to successfully extract oil from shale.

However, numerous problems hampered the company's progress. Stock sales were deficient, providing no more than one-half of the capital needed for a commercial plant to extract oil from shale. The shortage of funds precluded the construction of a two-mile aerial tram to carry shale from a mine to the processing plant. A continuous-process retort, designed by J. N. Wingett, was built and tested in Denver, but the company did not have sufficient funds to install it at the plant site in western Colorado. Dishonesty on the part of managers, in Denver and in the field, further eroded the enterprise's finances.

After two years, officials suspended operations. Reorganized in 1920 as a subsidiary of the Troy-American Petroleum Corporation, the firm experienced no relief from financial problems. The parent company was declared defunct in 1926, by which time all of the Colorado assets had been sold.

Several corporations operated successfully, usually on a small scale, in the Piceance Creek Basin. The Oil Shale Company built a plant on Dry Fork, twenty miles northeast of De Beque. Operations commenced in 1916 and ended in 1927. In all, the firm produced about 150 barrels, or 6,300 gallons, of oil, which it used locally for experiments in making soaps, ointments, and dips for livestock. Forty-two gallons constituted a barrel of oil, whether pumped from the ground or distilled from shale.

The Monarch Shale Oil Company, a Denver corporation headed by mining engineer Joseph H. Ginet, built a plant on Conn Creek, twelve miles northwest of De Beque. The enterprise utilized the so-called Ginet retort to produce about seventy-five barrels of oil. The Washington Shale Company, reorganized later as the Washington Shale Oil and Products Company, composed mainly of Seattle investors, struggled to develop an efficient retort. Its plant near Grand Valley produced small quantities of oil from shale. A continuous retort, installed in 1927, did not save the corporation from extinction, the fate suffered by other producers of oil from shale, because the cost was not competitive with crude oil pumped from underground reservoirs.

The Index Shale Oil Company had a slow start because of difficulty raising capital and finding an efficient retort. The firm, headed by Chester Church, a New York businessman, eventually adopted a retort invented by Harry L. Brown, a New Jersey entrepreneur. Some of the oil was used experimentally to make gasoline, what Church called "Indexoline." Operations ended in 1927, and the corporation was dissolved the following year.

Another corporation produced oil from Colorado shale, but not within the borders of the state. The Colorado Carbon Company, organized in November 1915, remained in operation for nine years. Under the direction of Lewis Leach, former mayor of Grand Valley, the firm mined and shipped shale to Denver and to Kansas City, Missouri, for processing. The kerogen was used to manufacture carbon black and wax.

By far the most successful American oil-shale enterprise of the 1920s was located in Nevada. Robert Catlin, an officer of the New Jersey Zinc Company, built an experimental plant in Elko to test a variety of processes for extracting oil from shale. The Catlin Shale Products Company, following its formation in 1919, functioned for five years, during which time it

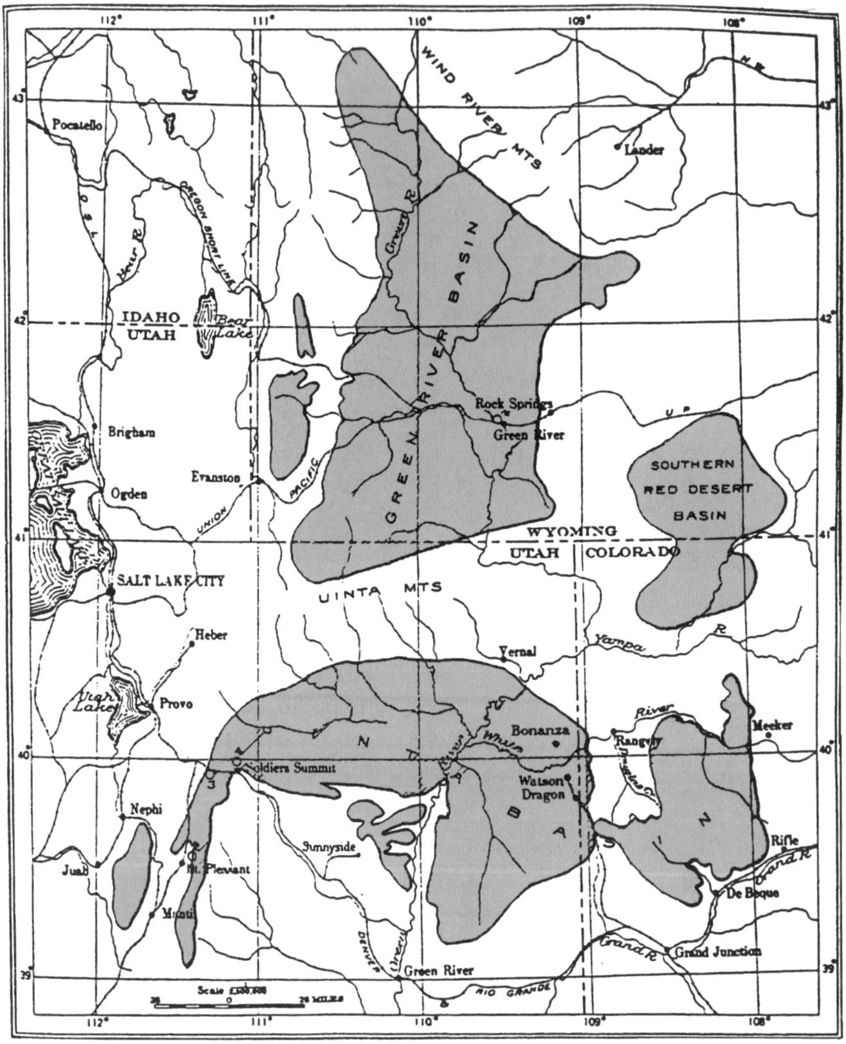

Map 4. Oil Shale Deposits in Colorado, Utah, and Wyoming. From U.S. Geological Survey, *Bulletin 641* (1917).

produced 12,000 barrels of oil. Catlin deserved to be called the "father of oil shale," but that appellation was bestowed on Victor C. Alderson, who did not mine or process so much as a single pound of shale.

A native of Massachusetts, Alderson was named president of the Colorado School of Mines in 1903. Ten years later he was forced to resign, allegedly for supporting students in a long-running quarrel with the faculty over the rigors of the instructional program. Reinstated in 1917, he served a second term as tumultuous as the first. He resigned again in 1925.

In 1917 the School of Mines launched a vigorous research program that focused on techniques of mining and extracting oil from shale. Alderson became a self-appointed national publicist for the new industry then taking shape in western Colorado. His unbounded, often unfounded, enthusiasm for oil shale was expressed in articles and interviews in newspapers and journals throughout the United States. He warned repeatedly that the business of extracting oil from shale did not have a bright future unless the volume of oil from wells failed to keep pace with rising demand for petroleum products. In addition, Alderson cautioned would-be investors that the processing of oil from shale, if cost-effective, would be a capital-intensive industry. He suggested that a successful plant in 1918 would have to handle 500 tons of shale a day, and that its cost would be a minimum of $150,000.

In the postwar years, widespread concern that the nation faced a shortage of crude petroleum proved to be unfounded. New fields, particularly in California, Oklahoma, and Texas, created a flood of oil that exceeded the needs of an expanding market. Abundant and cheap oil undermined the economics of the private oil-from-shale industry and hastened its demise in the 1920s.

Many people who were active in the petroleum industry understood that while the commercial potential of oil shale was in the distant future, there was a current need to explore the technology that could someday sustain systematic development. In addition, some Washington bureaucrats suggested that since the federal government was the principal landowner in the West, it should encourage experiments with new mining techniques, as well as the design and operation of cost-effective retorts. Reflecting this view, the Bureau of Mines prevailed upon Congress to authorize research studies on the technical aspects of deriving oil from shale. Two laboratories—one at the University of Colorado in Boulder, the other in Salt Lake City—pursued numerous studies until 1925. By that time,

Congress agreed to fund the Bureau of Mines' on-site experiments with oil shale.

Congress, largely at the insistence of Colorado's delegation, approved funding in 1924 to build and operate an experimental oil-shale plant. Representative Edward Taylor, from the Western Slope, and Senator Lawrence Phipps of Denver were primarily responsible for the legislation. President Calvin Coolidge signed an act in March 1925 that authorized the Bureau of Mines to test methods of mining and techniques for processing shale capable of producing oil for the United States Navy in the event of a future national emergency.

A site-selection committee examined portions of the Piceance Creek Basin near Grand Valley, De Beque, and Rifle before selecting a tract near Rulison, eleven miles west of Rifle. It was chosen largely because it afforded easy access to the reserve, which President Woodrow Wilson had created by setting aside a large tract of shale deposits in Colorado for the exclusive future use of the navy. In 1924, President Coolidge added 45,000 acres to the reserve. Rulison was located on the Colorado River adjacent to Navy Reserve Number Three. Water was readily available, as was transportation, for Rulison was on the main line of the Denver and Rio Grande Western Railroad.

The bureau approved the site in September, after which construction began on a structure to house two retorts and a variety of other buildings. The latter included bunk- and cookhouses, a combination office and laboratory, a powerhouse for generating electricity, and a machine shop. An aerial tramway transported shale from a mine located more than a mile distant from and nearly a quarter of a mile above the plant site. The tramway was the key to quarrying shale and delivering it to the plant at nominal cost.

The first attempt to extract oil from shale utilized a small model of the reliable Pumpherston retort, which had been used successfully for decades in Scotland. This was a so-called batch unit in that it processed 5 tons of ore at a time. Heat was applied externally to raise the temperature of shale to about 900 degrees Fahrenheit.

A second unit, a New York-Texas-Utah (N-T-U) retort, was installed at Rulison in 1926. The N-T-U was one of the more remarkable advances that emerged from the oil-shale boom of the World War I era. The prototype was designed and built by George Wallace, a St. Louis engineer. It stimulated development of new, more advanced designs in the 1920s because of its unique features. Unlike most of the so-called horizontal batch retorts of the time, Wallace's vertical unit was a continuous-run model with internal heating from combustion of gases and the carbon of spent shale to create the requisite temperatures for releasing kerogen from shale. Only by means of continuous operation, with heat derived from shale, could processors of oil from shale hope to achieve volume output at a cost that would allow competition with petroleum.

The most advanced retort of the time, the N-T-U suffered, nonetheless, a major shortcoming in that internal combustion consumed a large percentage of the kerogen that might otherwise have been produced from shale. The N-T-U Company of New York City was the distributor of the retorts, and it was from that firm that the Bureau of Mines purchased its unit for tests at Rulison. As the most promising retort of its time, it was the technology that engineers and technicians in Rulison hoped to improve upon in their quest for commercial production of oil from shale.

Experiments with retorts continued at the Rulison plant for a period of nearly three years. The plant closed on July 1, 1927, after all funds had been exhausted. A supplemental appropriation permitted the Bureau of Mines to reactivate the plant in 1928 and to keep it open for an additional fiscal year.

Operations at Rulison revealed that a large amount of oil could be extracted from shale. The plant processed more than 6,000 tons of shale to produce 3,568 barrels of oil. This amount of oil was, in fact, approximately the goal that had been set for the project in 1924. Local officials objected to a quota, insisting that the goal should not be a specific amount of oil, but a methodical test of equipment and techniques for mining and processing shale. The champion of this view, superintendent Martin J. Gavin, did not return to Rulison when work resumed in 1928. Boyd Guthrie, Gavin's former assistant, was the on-site supervisor during the final year of operations.

The experiments in western Colorado revealed that numerous problems remained to be solved before there could be commercial production of oil from shale. While the Rulison project proved beyond question that oil could be extracted from the shale of the Green River formation, the cost of doing so ruled out competition with petroleum. Furthermore, the quality of the product was questionable. Oil from shale was used to make gasoline and fuel oil, but refining costs were greater than with crude petroleum. For this reason, refiners did not view oil from shale as an ideal substitute for crude oil.

Despite what should have been seen as obvious limitations to the technology for removing oil from shale, boosters hailed the bureau's Rulison program as an achievement that indicated the time for commercial production of oil from shale was at hand. More cautious observers conceded that oil shale was a potentially valuable resource that might be commercially developed at some future date when petroleum was no longer plentiful and cheap. However, by the time this view was promulgated, the Rulison project had ended. By the mid-1920s, the supply of crude oil exceeded demand, and there seemed to be little possibility that production from underground reservoirs would diminish in the foreseeable future.

In fact, throughout the Depression years of the 1930s, the federal government and the leading oil-producing states worked together to curb the oversupply of petroleum. Not until World War II did the demand catch up with and threaten to exceed supply. By 1943, due to wartime conditions, the United States once more faced the possibility of an oil shortage.

Since the nation's war machine ran on petroleum, government officials had to look for large-scale alternative sources. The initiative for the development of what came to be called a synthetic liquid-fuels program came from the Petroleum Administration for War, headed by Secretary of the Interior Harold Ickes. Politicians from leading coal-mining states, especially Senator Joseph C. O'Mahoney of Wyoming and Representative Jenning Randolph of West Virginia, guided through Congress what became the Synthetic Liquid Fuels Act of April 5, 1944. Approved overwhelmingly by Congress and signed by President Franklin D. Roosevelt, this legislation sought to stimulate the commercial production of oil from new sources in order to aid the prosecution of the war while at the same time conserving the nation's limited natural petroleum resources.

The Department of the Interior, acting through the Bureau of Mines, was authorized to construct and operate plants to develop techniques for the production of liquid fuels from coal, oil shale, and other substances. Cost and engineering data gathered by the bureau from the operation of experimental plants were to be distributed to private enterprises in order to encourage the development of permanent synthetic liquid-fuels industries. Congress extended the program in 1948 with additional funding, giving it a life span of eleven years at a total cost of about $87 million.

Under the authority of the act of 1944, the Bureau of Mines established the Office of Liquid Fuels in Washington, D.C., to oversee several projects. These included coal-to-oil experimental research stations in Pennsylvania, West Virginia, and Missouri. An Oil-Shale Experiment Station was located at Anvil Points, near Rifle, and an Oil-Shale Research Laboratory was established at Laramie, Wyoming. While all of the projects had the potential of affecting the intermountain states, only the extraction of oil from shale had a direct impact on the region.

The Office of Liquid Fuels established an oil-shale experimental site in the Piceance Creek Basin in Garfield County, Colorado. A selection team led by Boyd Guthrie, who was to serve as superintendent of the project, examined the area before choosing a site called Anvil Points, a name taken from the silhouettes of nearby mountains, seven miles west of Rifle. Construction commenced immediately on roads, a mine, a retort building, a refinery, water and sanitation systems, an electric plant, and houses for personnel.

The Anvil Points plant, in conjunction with the Oil-Shale Research Laboratory in Laramie, hoped to develop efficient methods of mining shale as the first step toward making the extraction of oil from shale commercially feasible. Mining techniques, designed to yield a large volume of shale at low cost, fulfilled all expectations. A large mine was excavated from a steep cliff 2,200 feet above, and via a switchback road, five and a half miles from the plant. Shale was mined by mechanical methods in a cavern with a 70-foot ceiling. Extraction and transportation were highly efficient. The cost of ore for the retorts was about 50 cents a ton, low enough to sustain future commercial operations.

The extraction of oil from shale began in May 1947, using two 40-ton N-T-U batch retorts. As anticipated, they were unsatisfactory because of small size. In addition, they required water for cooling, and the volume of oil recovered was small. Although obsolete, the N-T-Us provided valuable technical data on internal-combustion processes, and produced oil without delay for refining tests.

The commercial production of oil from shale required an efficient, low-cost, high-capacity process. The ideal retort had to be thermally self-sufficient, with continuous high rates of feed, efficient oil recovery, and requiring little or no water consumption. It had to be mechanically simple, expandable for large-volume operations, and economical in terms of original investment and operating costs. And it had to produce oil at costs that made it competitive with natural petroleum.

Great strides were made toward the achievement of these goals. A 6-ton-capacity-per-day Gas-Combustion-Retort was followed by larger models with capacities of 25 and 150 tons. These were continuous gravity-feed retorts. Vertical vessels of circular cross section, they were internally heated by hot, recycled retort gases obtained from the combustion of residual carbon in the spent shale. In the larger models, crushed shale moved downward against a countercurrent of hot gases. An outstanding feature of the Gas-Combustion-Retort was that water was not required for cooling the retort or for the oil-recovery system. This was important because the shale deposits of the Green River formation were in a region where water was always scarce and often nonexistent. The goal of an efficient 300-ton retort appeared feasible at the time operations ended in July 1955.

Petroleum products refined from oil extracted from shale at Anvil Points were subjected to a variety of tests. One, designed to gain recognition and support for the experimental plant, involved the operation of the Denver and Rio Grande Western's passenger train, the "Prospector," on oil-from-shale during its regularly scheduled run from Salt Lake City to Denver on September 1, 1950. For that purpose, the locomotives' fuel tanks were filled with 3,750 gallons of diesel refined from 624 tons of shale. In addition, that quantity of shale yielded an equal amount of gasoline, 6,600 gallons of heavy fuel oil, almost 7 tons of petroleum coke, plus other by-products.

In spite of widely publicized achievements in developing mining methods and retorting technologies, funding for activities at Anvil Points ceased in the mid-1950s. The petroleum industry viewed the project as an unnecessary public enterprise. President Dwight D. Eisenhower and Congress concurred. Congress's original authorization for a five-year program had been twice extended, and the funding had more than doubled in the course of the project. The Eisenhower administration urged Congress to terminate funding on the grounds that it was time for private enterprise to carry forward the commercial processing of oil shale. Critics charged that Eisenhower bowed to the demands of the petroleum industry, which feared the potential competition of oil from shale.

A vocal opponent of closing the Anvil Points plant was Warwick M. Downing, a Denver independent oilman. Born in Macomb, Illinois, in 1875, Downing moved to Denver as a young man and became a permanent resident after earning a law degree at the University of Michigan. A specialist in mining law, Downing actively participated in the petroleum industry beginning in 1906 as an investor in a well drilled in the Boulder field. He later played a role in the development of several fields in Wyoming. He represented Colorado on the Oil States Advisory Committee from 1931 to 1934, and participated in the administration of the Interstate Compact as Colorado's representative on the Gas Conservation Commission from 1935 to 1951. He was a member of the National Petroleum Council (NPC), a powerful trade association composed largely of representatives of major oil companies. He was chairman of the Colorado Oil and Gas Conservation Commission from its inception in 1951 until shortly before his death in 1964.

Downing had been an early advocate of oil-shale development. His public plea, in the midst of an apparent oil shortage in 1943, appears to have been instrumental in mobilizing the support of congressional delegations from western states for the Synthetic Liquid Fuels Act of 1944. A decade later he mobilized support in Colorado, in adjacent states, and in Washington, D.C., for continuing the Bureau of Mines' experiments at Anvil Points.

According to Downing, Secretary of the Interior Douglas McKay acted against the best interest of the nation by insisting that the experimental station in Colorado had to be closed unless the petroleum industry was willing to take over the project. McKay placed the fate of the Anvil Points facility in the hands of the NPC with assurance of the outcome, for seventeen of the twenty-four members were presidents or upper-level managers of the nation's major oil enterprises, many of whom had a vested interest in perpetuating America's growing dependence on Middle Eastern oil. Downing believed that a majority of the NPC viewed the Rocky Mountain oil-shale industry as a threat to imports, and it was for that reason the council favored ending the government's oil-shale investigations.

Rather than accept the Eisenhower administration's decision, Downing took his quarrel with the executive branch to the Congress. The delegations from western states united in a bipartisan effort to block the closing of Anvil Points. In 1955, Wyoming's O'Mahoney, along with Senator Gordon Allott of Colorado, gained additional funding and a one-year reprieve. Although it was scheduled to end in April 1955, Congress funded aspects of the program until mid-1956. Thereafter, the federal government's technical investigations were confined to activities by the Bureau of Mines' Petroleum Research Center in Laramie, while the Anvil Points facility was placed on standby status pending future need for oil from shale.

Unwilling to give up the potential of oil from shale in the event of a national emergency, the navy sought unsuccessfully to underwrite continued experiments at Anvil Points. In June 1956, Captain A. S. Miller, director of Naval Petroleum and Oil Shale Reserves, at the request of the Department of the Interior and after gaining the approval of the House Armed Services Committee, took possession of the oil-shale plant and developed plans for its continued operation. However, Congress refused to approve funds for experiments, and the navy withdrew, leaving the Anvil Points facility on standby status and in the possession of the Department of the Interior.

The closing of the government facility did not end investigations into the potential of oil shale for commercial development. Two corporations—the Union Oil Company of California and Sinclair Research Inc.—were active by the early 1950s. Union Oil Company of California (later renamed UNOCAL Corporation) had a long involvement with oil shale. That firm acquired some 30,000 acres in the Piceance Creek Basin in the 1920s, but did not attempt to develop that resource because petroleum production sharply increased with the discovery of new fields, particularly in Texas and Oklahoma. In 1943, in the midst of World War II, when another oil shortage seemed imminent, Union Oil resumed shale investigations by building and operating a small retort in California. Again, increased production of cheap petroleum discouraged a long-term effort to produce oil from shale.

A decade after the close of World War II, Union Oil returned to the Piceance Creek Basin. On May 18, 1957, the company dedicated a Shale Demonstration Plant and a mine near Parachute. By 1958, the corporation's vice-president for research announced the plant had been in continuous operation twenty-four hours a day for as much as six weeks at a time, during which shale had been processed at a rate of more than 1,000 tons a day. The plant had a rated capacity of 1,000 barrels of oil per day and shale was mined at a cost of less than $1 a ton. Despite this technical success, the plant was shut down because petroleum was plentiful and cheap.

A different approach for extracting oil from shale was initiated by the Sinclair Oil and Gas Company through its subsidiary, Sinclair Research Inc. At a test site in the southern section of the Piceance Creek Basin, engineers and technicians sought to determine the feasibility of recovery of oil from shale in situ. This required drilling holes, the application of hydraulic pressure to fracture underground shale deposits, the use of heat to release kerogen, and pumps to bring it to the surface. Sinclair's engineers successfully tested small-scale, in situ retortings in 1953–1954, and again in the 1960s. Although the company did not publicly report its experiments at the time, it seems likely that there was not enough fracturing to ensure the commercial success of the in situ method of extracting oil from shale.

Promising investigations with in situ extraction occurred under the auspices of the Bureau of Mines near Rock Springs, Wyoming, in the 1960s and early 1970s. Several wells were drilled within a compact area to gain access to a stratum of oil-rich shale at depths of 68 to 88 feet below the surface of the ground. The deposits were fractured by means of high-voltage electricity, pressurized water, and the detonation of liquid explosives. Combustion was initiated with a propane burner and sustained by forcing air into the central ignition zone. About 190 barrels of oil were extracted for analysis from wells at the rate of 4.5 barrels per day. The bureau's personnel concluded from the tests that more powerful explosives, particularly nuclear devices, could greatly enhance the efficiency of the in situ process.

The possibilities were breathtaking. One explosion could theoretically create a huge underground cavity filled with broken shale. With the application of heat, as much as a million barrels of oil might be obtained from a single site. This method of rock fracturing, first proposed to enhance natural-gas production, appeared to be ideal for large-scale, in situ production of oil from shale. The Atomic Energy Commission and Bureau of Mines enlisted the support of private enterprises for a test of nuclear energy's potential for underground, in-place retorting. The American Petroleum Institute actively supported the project. However, this scheme was temporarily dropped when the United States voluntarily ended nuclear testing in 1959.

The project was revived in the mid-1960s, but the goal was to enhance the production of natural gas, not oil from shale. Although the fledgling oil-shale industry was not directly involved in large-scale fracturing of underground structures, it was concerned that the use of nuclear detonations to free natural gas from tightly compacted rock formations would jeopardize the extraction of oil from shale. For that reason, participants in the oil-from-shale industry looked with disfavor on joint efforts by government agencies and private enterprise involving nuclear stimulation of natural-gas production.

Furthermore, an underground nuclear explosion was seen as a threat to the deposits in the Piceance Creek Basin at a time when commercial development appeared to be near at hand. Local enterprises, led by The Oil Shale Company (TOSCO), did not oppose the stimulation of underground gas as such, only the use of nuclear energy if it endangered the extraction of oil from shale. They believed that oil-shale development was the only sure way to augment America's dwindling energy supplies.

Opponents of the nuclear blasts, including the oil-shale advocates, mobilized to block Project Rio Blanco. The public outcry, combined with technical difficulties of utilizing natural gas produced by nuclear stimulation, ended the Atomic Energy Commission's plans for further stimulation of gas production. This eliminated any threat that nuclear explosions might have posed to the shale resources of the Piceance Creek Basin.

The cause of oil shale received a big boost in 1964, when the Colorado School of Mines Foundation, in behalf of several major oil companies, leased the plant at Anvil Points. The objective was to seek advances in mining and retorting techniques developed a decade earlier by the Bureau of Mines. Participants included the project manager, Mobil Oil Company, plus Humble Oil and Refining Company, Continental Oil Company, Pan American Petroleum Corporation, Phillips Petroleum Company, and Sinclair Research Inc. Over a period of five years, the partnership invested approximately $4.5 million in mining and retorting experiments. Shale mining was successful in that a large volume was produced at nominal costs. Less satisfactory experiments with retorting techniques undermined hopes for immediate commercial development of oil-shale resources. Operations ended in April 1968.

Similar tests were conducted by the Colony Development Company in the Piceance Creek Basin. Organized in April 1964, it served as agent for TOSCO, Standard Oil of Ohio (Sohio), and the Cleveland Cliffs Iron Company. Colony erected a pilot plant seventeen miles north of Parachute to test TOSCO's process for extracting oil from shale.

TOSCO's founder, H. E. Linden, a California businessman, purchased the American rights to a method of extracting oil from shale that had been devised by Aspengre and Company of Sweden. Called the Aspeco process, it utilized heated ceramic balls in a revolving retort to vaporize the kerogen in shale. In 1967, the TOSCO II process, a variation on Aspeco developed by Linden, was tested at a plant erected near Littleton, Colorado. These experiments were conducted by the Denver Research Institute, a division of the University of Denver.

Operations began at Colony's plant in western Colorado in mid-1965. The Cleveland Cliffs company handled mining operations. TOSCO operated a large retort with a rated capacity of 1,000 tons of shale. By 1967, this unit was producing about 1,000 barrels of oil a day. Three years later, Morton M. Winston, executive vice-president, told a congressional committee that TOSCO had solved all technical problems related to the production of oil by the Aspeco process and could, if the market price of oil justified an effort, turn out 58,000 barrels per day.

What looked like the start of another oil-shale boom was taking shape in western Colorado by the 1960s. This time, the major oil companies were important participants, for they had acquired most of the privately held shale lands in the Green River formation, particularly in the Piceance Creek Basin. For nearly half a century, the only people who made money from oil shale were sellers of patented claims. Prior to 1920, when it was easy to obtain patents on claims, Piceance Creek Basin shale land sold for about 75 cents per acre. The Minerals Leasing Act of 1920 ended the practice of filing claims to secure patents and titles by authorizing only leases from the federal government. Thereafter, prices for private land increased from approximately $5 to as much as $75 an acre by the close of the 1960s.

The buyers were principally producers and distributors of crude petroleum and oil products. By the 1960s, private holdings in the Piceance Creek Basin exceeded 223,000 acres, two-thirds of which were owned by ten major oil companies. Standard Oil of California, Union Oil of California, Getty Oil, and the Texas Company had the largest holdings. In addition, Union Oil owned most of the farms and ranches along Parachute Creek for a distance of twelve miles.

Private holdings in the Piceance Creek Basin were only a small fraction of the area's oil-shale lands, and by no means the richest deposits. The federal government owned more than 70 percent of the oil-shale land and 80 percent of the richest deposits. The potential of this land as a source of oil at a time when domestic production failed to keep pace with demand convinced the Department of the Interior, particularly Undersecretary Morris Udall, that government policy should encourage private enterprises to take up leases for experimental development that could lead to commercial production.

The announcement of plans to lease the nation's shale deposits raised the specter of a land grab by the major oil companies in an effort to perpetuate their monopoly on petroleum production and distribution. Critics of the Interior Department's proposal denounced it as a giveaway to corporate America. J. R. Freeman, a fiery Firestone, Colorado, newspaper editor, became the unlikely leader of a mass protest to forestall the "rape" of the country's petroleum reserve.

The outcry forced the Department of the Interior to modify the leases. New terms called for strict environmental controls and higher royalties. In addition, the agency offered only three lease sites in western Colorado. This new stance cooled industry's enthusiasm for on-site research and development. On December 20, 1968, three bids, with a combined value of slightly more than $500,000, were submitted for 11,463 acres containing an estimated 2.3 billion barrels of oil. The department rejected them as inadequate.

However, in 1971 a sharp decline in domestic oil production prompted a decision to again offer tracts in the Green River formation, specifically two each in Colorado, Utah, and Wyoming. These properties were not put up for auction until 1974, by which time an Arab oil embargo had intensified the potential for energy shortages in the United States. This had the effect of stimulating active interest in oil-shale development. Another boom was under way, and this time it seemed almost certain that there would be commercial development.

The likely decline of the domestic petroleum industry focused attention on more than oil shale. It highlighted, in addition, the potential for deriving oil from coal. Industrial liquefaction, as well as gasification, of coal had been highly developed in the nineteenth century. Needed were new technologies that would be both cost-effective and capable of producing high-quality oil and gas from America's most abundant energy resource, solid fossil fuel.

Systematic research in coal liquefaction and gasification had begun with the Synthetic Liquid Fuels Act of 1944. It authorized the construction and operation of demonstration plants to produce synthetic fuels from coal, oil shales, agricultural and forestry products, and other substances. Demonstration facilities for converting coal to liquid fuels were erected at Louisiana, in eastern Missouri, at what had been a wartime plant for making synthetic ammonia.

The Synthetic Liquid Fuels Act of 1944, as amended in 1948, authorized other research activities that were unrelated to coal. A portion of the initial funding underwrote a four-year program conducted by the Department of Agriculture to construct and operate a plant for producing ethanol from sugar and corn cobs. Some attention was directed, at the same time, to secondary recovery techniques for stripper wells and new refining processes.

This work was conducted from 1948 to 1950, principally at Bartlesville, Oklahoma, and Laramie, Wyoming.

Coal demonstration plants remained active until 1955. After an intense investigation of processes for making oil and oil products from coal, it was clear that technical problems remained to be solved before synthetic fuels could be competitive with petroleum. Furthermore, by 1955 a worldwide petroleum surplus undermined any rationale for large-scale conversion of coal to oil, leading officials of the Bureau of Mines to conclude that further studies should be confined to intensive laboratory research and small pilot plants. Therefore, the synthetic liquid-fuel projects were absorbed into the bureau's regular research programs, which emphasized efficient and safe methods for mining coal.

At the insistence of congressmen who wanted research projects for their states, the Office of Coal Research (OCR) devoted most of its limited budget to finding new ways of utilizing coal for the development of synthetic fuels. Eight major projects were approved, two of which were terminated in 1967–1968 because of inadequate funds. The other six were divided evenly between coal liquefaction and coal gasification. In 1968, 86 percent of the federal money budgeted for research was for "synfuels" pilot plants. Only two of the principal projects were located west of the Mississippi River.

OCR funded small but significant projects in the intermountain states. A contract was awarded in 1968 to the University of Wyoming's Natural Resources Research Institute for a five-year study of methods for converting coal to gasoline and other petroleum products. Along that same line of research, the School of Mineral Industries, University of Utah, prepared a report summarizing laboratory investigations of various processes for converting coal into liquid fuels.

By the close of the 1960s, the federal government's OCR program for the development of synthetic fuels was in deep trouble and seemed destined for failure. It floundered because it lacked broad-based support either in the Congress or in the executive branch of government. This was compounded by private industry's hands-off attitude. Given the assumption of surpluses of both petroleum and natural gas, major energy companies were reluctant to make substantial investments in developing commercial facilities for producing synthetic fuels from coal.

Ironically, at the time that failure seemed certain, a threatened energy shortage stimulated new government initiatives for synfuels development.

In the 1970s, the decline of domestic oil and gas reserves, at a time when consumption expanded at a rapid pace, prompted Congress and private industry to change their perception of synfuels. With increased funding and the cooperation of oil and gas companies, OCR underwrote pilot plants, three liquefaction and five gasification projects, plus numerous research efforts designed to develop new synthetic-fuel technologies. Most of the pilot plants were successful in producing data needed for the development of demonstration plants. Additional projects were suggested but did not achieve fruition. Intermountain West states with large strippable coal reserves were identified as tentative sites for synfuel plants.

This casual approach to synthetic fuels changed to one of urgency in the autumn of 1973 when the Arab oil producers, in response to American military assistance to Israel during the Yom Kippur War, imposed a partial embargo on exports to the United States. The reduction in the flow of oil from the Middle East exacerbated a growing imbalance between energy demand and supply. The emergence of what was popularly called the "energy crisis" changed the public's attitude toward synfuels development. It was no longer a question of whether there should be commercial extraction of oil and gas from coal, but how soon and in what manner a program could be implemented.

9

Energy Crisis 1973-1985

The demand for energy within the United States and worldwide had been closing in on available supplies since the end of World War II. However, most Americans were unaware of a pending shortage, or that it was approaching crisis proportions, until an interruption in imports sent oil prices skyrocketing. The end of abundant cheap energy threatened to undermine a highly materialistic lifestyle based on large-volume, often wasteful, consumption of fossil fuels. A shortage of energy—especially oil—and its spiraling cost altered, at least in the short run, social as well as economic behavior and forced people to make what were often painful adjustments to a new reality of high-cost energy. For a time, the nation's resolve to regain energy self-sufficiency resulted in efforts to expand domestic production of coal, oil, and natural gas. Simultaneously, a variety of new technologies were explored as means of producing synthetic and alternative fuels.

What most people called an "energy crisis" seemed to suddenly materialize in October 1973, when representatives of Arab oil-producing nations, meeting in Kuwait, announced a sharp increase in the price of crude and a reduction of 5 percent a month in the export of oil until Israel, then caught up in the Yom Kippur War, withdrew from all territory taken from its Arab neighbors. This announcement was quickly superseded by a total

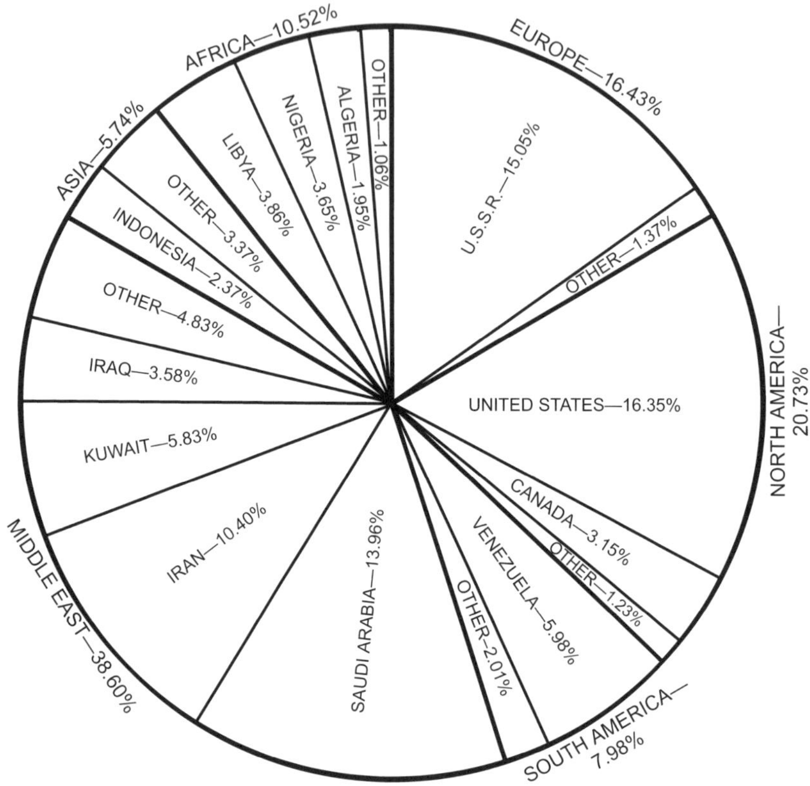

Figure 8. World Share of Crude Oil Production, 1973—20.56 Billion Barrels. From U.S. Bureau of Mines, *Minerals Year Book, 1973* (1976).

embargo of oil shipments to the United States and the Netherlands, the two nations that, in the Arab view, were the most consistent supporters of Israel's efforts to build a nation in the area the Palestinians called their homeland.

At the time, the United States led all nations in the output of crude oil, as shown in Figure 8. For this reason, most Americans could not understand why an oil embargo triggered a motor-fuel shortage in the United States. It was incomprehensible. Even petroleum-market analysts were decidedly ambivalent about the causes of the crisis. Some insisted that the shortage was real, and that it had been evolving over a period of more than a decade. Others concluded that the embargo, effective from October 1973 to March 1974, allowed major oil companies to turn an alleged shortage into an opportunity to advance prices and to enhance corporate profits.

Real or unreal, the crisis escalated fuel prices, stimulated inflation, and threatened to undermine the vitality of the American economy. For the balance of the 1970s, the United States was in the grip of "stagflation," a malaise of simultaneous inflation and economic stagnation. Because the traditional remedy for one was the deliberate pursuit of the other, the nation could not extricate itself from the duality of rising prices and slowing business activity until energy prices, particularly oil, started downward in the 1980s.

For most Americans, the shortage of oil had its greatest impact at the gasoline pump. A highly mobile way of life based on private ownership and operation of vehicles powered by internal-combustion engines was placed in jeopardy. Motorists had to line up, often for hours, in the hope of purchasing at best a few gallons of gasoline. The longer the wait, the shorter the tempers. Patience exhausted, people asked why the United States was not capable of meeting its own fuel needs. They could not understand why the world's most powerful nation appeared to be held hostage by a few Middle Eastern sheikdoms due to its support of Israel.

President Richard M. Nixon and his advisors assumed that at the heart of the problem was America's increasing dependence on foreign oil. Responding to mounting public anger, Nixon proposed that the nation increase domestic output of all sources of energy and at the same time adopt a program of conservation to reduce consumption. Project Independence, outlined in a televised presidential address on November 25, 1973, called for a national commitment to energy self-sufficiency by 1980. This required, Nixon insisted, increased production of coal, oil, natural gas, and synthetic fuels, principally oil from shale, plus oil and gas from coal. In the address, Nixon also called for the expansion of electricity-generating facilities.

The president recommended to Congress specific actions for expanding energy output. Included was a call for more nuclear-powered generating facilities and an accelerated program of leasing federal offshore lands for oil exploration. He also proposed a ban on conversions from coal to oil or gas by public utilities and other enterprises, and reductions in the government's allocation of fuel for commercial air travel.

Furthermore, Nixon asked the American people to adopt voluntary programs for conserving energy. Homeowners were to set daytime thermostats during the heating months at 68 degrees Fahrenheit. Managers of businesses were to reduce energy consumption by 10 percent. States were to reduce the consumption of gasoline by restricting highway speeds of automobiles and other vehicles to 55 miles an hour.

Other suggestions called for the closing of gasoline service stations on Sundays, the banning of outdoor lighting for advertising or displays, and restrictions on the production and distribution of gasoline. Nixon stated that he was prepared to ration gasoline if Congress authorized that action. In fact, he already had the power to do so, but preferred that the initiative for that potentially unpopular action come from the legislative branch of government.

The goal of energy self-sufficiency by 1980 was unrealistic, yet a public commitment was a political necessity. With these measures Nixon hoped to increase supply while decreasing the demand for energy. Output was projected to expand at a rate of 4.7 percent per year, while consumption was projected to decline initially at a rate of 3.6 percent annually, adjusted to 2 percent before the close of the decade.

Sharp increases in the supplies of all fossil fuels were required. Coal output was projected to expand by 60 percent from 1973 to 1980. At the same time, the production of crude oil was to grow from 10.9 to 14 million barrels a day. The yield of natural gas was to increase to 4 trillion cubic feet a year. These were commendable goals, but it was not likely that the energy industry could achieve all of them.

The producers of coal were asked to assume the principal burden for achieving national self-sufficiency. The primary stimulus for expanded output was to be elimination of federal price controls. Other incentives were to encourage the rapid expansion of oil and gas, and the development of synthetic fuels by means of the liquefaction and gasification of the nation's abundant solid-fuel resources. In order to hasten development, some national air-quality standards were to be set aside for up to five years. Also needed was an expanded coal transportation system and the conversion of oil- and gas-burning industrial plants to coal.

Nixon's successors, Presidents Gerald Ford and Jimmy Carter, reaffirmed the goals of Project Independence, but moved back the target date for national energy self-sufficiency to 1985. The goal was to be achieved primarily by expanding the production and utilization of coal and nuclear power–generated electricity. The output of solid fossil fuel was to be increased by 1985 to the point that it would eliminate the need for petroleum imports in the amount of more than 8 billion barrels a year, with annual savings to the nation of about $10 billion.

The emphasis was on increased output of domestic energy supplies. In Ford's plan, coal production in the West was to expand at the rate of 250 million tons a year, much of it to be consumed by new coal-fired, steam-generating electric power plants or converted to synthetic oil and gas. Both Nixon and Ford recommended modifications in surface-mining laws, the easing of clean-air restrictions, and the simplifying of power-plant location procedures to encourage local production and consumption of coal. Carter endorsed increased coal output, but also emphasized the expansion of federally supported research and development programs devoted to coal mining, combustion efficiency, and synthetic-fuel technologies.

During the 1970s, Congress, at the urging of three successive presidents, approved several laws designed to achieve the goal of national energy self-sufficiency. The Energy Supply and Environmental Coordination Act of 1974 authorized the Federal Energy Administration to require power plants and other large fuel consumers to convert from oil and natural gas to coal. The Energy Policy and Conservation Act bolstered the energy administration's authority to order conversions to coal for existing plants, as well as for those under construction. The National Energy Act of 1978 encouraged coal utilization by allowing the price of natural gas to rise through deregulation, and by prohibiting the use of oil and natural gas in new utility and industrial plants.

The impact of Project Independence fell heavily on the intermountain West, a regional storehouse of both nonfossil and fossil fuels. The former were abundant. Four states—Colorado, New Mexico, Wyoming, and Utah—contained about 95 percent of the nation's recoverable uranium, which was the fuel for nuclear electric-power generation. Geothermal energy resources were located in portions of Colorado, Utah, Wyoming, Montana, and Idaho. Most of the intermountain states had substantial reserves of petroleum, natural gas, and natural-gas liquids. Tar sands (asphalt rocks) were located principally in Utah, while that state, plus Colorado and Wyoming, contained rich deposits of oil shale.

Coal was by far the region's most abundant energy resource. In 1974, six states in the intermountain region contained almost half of the nation's known coal reserves, and most of it contained less than 3 percent sulfur. Furthermore, a large percentage of the nation's strippable solid-fuel deposits were in the region. In the West, as elsewhere, Project Independence stimulated coal production. Growth was particularly startling wherever large, shallow, relatively flat deposits of low-sulfur coal permitted the surface extraction of coal at costs significantly lower than from underground mines.

Rapid expansion was relatively easy in the West because vast reserves were located on federally administered lands, and to a large degree they were ideally suited for surface mining. The public lands were controlled by the Bureau of Land Management. In addition, most of the coal suitable for surface mining in Arizona and New Mexico, as well as in one area of Montana, was on Indian reservations. In 1960, production from federally administered lands amounted to 1.2 percent of the nation's output. Fifteen years later it was 30.2 percent of the total.

The market for coal had changed dramatically by the 1970s. No longer was it used extensively in the residential, commercial, and transportation sectors of the economy, and it could not compete with oil and gas in a large part of the industrial sector. Coal was widely consumed, however, for the generation of electricity. Responding to a booming economy in the 1950s, electric utilities expanded, erecting coal-fired, steam-generating power stations because the solid fossil fuel was more economical than oil or gas. By the 1970s, the utilities were struggling to keep pace with demand. The growth of housing and industry in ever-mushrooming suburbs required more and more electricity.

Much of the rising demand for electricity reflected the adoption of modern conveniences by Americans whose lifestyles were based on high-level energy consumption. Homes with electrical appliances were hallmarks of affluence. By 1980, more than four-fifths of all dwellings had one or more television sets. Microwave ovens were relatively new and not widely used, but 54 percent of homes had electric ranges. Clothes washers and dryers were found, respectively, in 75 and 47 percent of homes, and 37 percent featured dishwashers. Iceboxes had been superseded by refrigerators in 86 percent of homes, and 14 percent of homes had two or more units. Separate freezers were located in 38 percent of homes.

In addition, most homes built after World War II featured forced-air heating using natural gas or oil. This permitted the coupling of furnaces with air conditioning for central heating and cooling. A majority of dwellings were mechanically cooled in 1980, 27 percent with central systems and 30 percent with one or more individual room units. That left 43 percent of the homes in America that were without mechanical cooling devices of any kind.

The rapid expansion of the electric utility industry in Colorado called for increased production of coal. New mines were opened, and older properties expanded to supply fuel to power plants. After suffering through a fifty-year depression earlier in the century, coal mining enjoyed a welcome return of prosperity. Under the stimulus of Project Independence, the industry set new production records. In 1974, Colorado produced nearly 7 million tons of coal. By 1978, the state's mines surpassed the record output of 12.7 million tons set in 1918.

Surface mines, with lower costs than underground operations, were responsible for much of the expanded output. In 1974, seven surface operations accounted for 52.8 percent of the state's output, eclipsing the combined output of twenty-eight underground mines. Only two counties, Gunnison and Routt, recorded output in excess of 1 million tons; the former with 1.3 million tons, mostly for the steel plant in Utah, and the latter with 3.5 million tons, mostly for the region's electric-generating plants. Nearly 50 percent of the state's production came from Routt County's five mines; four surface operations accounted for more than 90 percent of the total. The Edna Mine, the Energy Mines One and Two, and the Seneca Two Mine had a combined output of 3.4 million tons.

Increased production, because of the bulk and weight of coal as well as the often remote location of mines, necessitated an efficient transportation system to move coal to markets. Historically, railroads had provided long-distance services, but the rapid expansion of output threatened to overwhelm the carriers. At the same time, rising freight rates forced mining enterprises to investigate alternative methods for carrying large volumes of coal to markets at reasonable cost. An attractive option, moving pulverized coal in combination with water through pipelines, had been successfully tested in Ohio in the late 1950s, but operation had been suspended when railroads lowered rates for moving coal from mines to markets.

In 1970 the successful movement of coal by slurry pipeline from Arizona's Black Mesa to the Mojave generating station at Davis, Nevada, focused attention on that mode of transportation as an alternative to railroads. Numerous pipelines were projected to carry western coal to distant markets during the 1970s and 1980s. Some plans were highly controversial, for example that of Energy Transportation Systems Inc., a Bechtel Corporation subsidiary, which planned to construct a billion-dollar pipeline to transport 25 million tons of coal per year from mines in Wyoming's Powder River Basin to White Bluff, Arkansas, and Baton Rouge, Louisiana.

This coal was to be consumed by electric-generating plants in the South. An aquifer underlying much of western South Dakota and eastern Wyoming was to provide the water to create the coal slurry.

Equally controversial was a proposal by the San Marco Pipeline Company to move 15 million tons of coal per year from mines near Walsenburg, Colorado, to Houston, Texas, for consumption by utilities. Water for this project was to come from Colorado's San Luis Valley. This and similar schemes faced countless challenges, the most serious of which was the acquisition of rights-of-way.

Railroads led the opposition because slurry pipelines threatened their lucrative monopoly on long-haul transportation of coal. Some environmental organizations looked with disfavor on the movement of coal through pipelines because that might lead to an expansion of the mining industry as well as higher consumption of the fuel by the electric-generating stations. Congress, responding to the fears of both groups, delivered a fatal blow to slurry schemes in 1978 when it refused to authorize the right of eminent domain in the public interest for projected lines. Without the power to secure rights-of-way by means of legal condemnation proceedings, the pipelines could not be built at reasonable cost.

As events turned out, the perpetuation of the railroad's de facto monopoly on long-distance transportation did not slow the growth of the fuel industry in the intermountain West. The Staggers law of 1976, more correctly known as the Railroad Revitalization and Regulatory Act, pumped new life into what had been seen as a moribund system. The carriers were allowed to abandon unprofitable routes, merge with other lines, and reduce costs by eliminating jobs, many of which had been made redundant by technological advances. In addition, rail lines were accorded the right to contract for services, including the provision of coal to utilities.

The carriers, responding to demands for lower coal freight tariffs, expanded the use of unit trains, which had evolved as a specialized form of transportation since the 1960s. Employed on a large scale to move approximately 10,000 tons of coal at a time from mines of the intermountain West to the Midwest and South, unit trains were also widely used to transport fuel from mines to electric utilities within the region. Nationally, electric power plants consumed two-thirds of all mined coal. The remainder was used in approximately equal amounts by coke plants and other industrial enterprises. Only about 1 percent of all mined coal was purchased by residential customers.

Colorado's output continued on an upward track until 1982, when forty-four mines produced a total of 18.5 million tons, down from 19.7 million the previous year. Sixteen surface mines produced 63.1 percent of the state's total. The surface operations in Routt and Moffat Counties accounted for 10.8 million tons, or well over half of all the coal mined in Colorado. The Colowyo Mine was the leader with 3.2 million tons, followed by the Energy One Mine with 2.9 million tons, the Trapper Mine with 2.0 million tons, and the Seneca Two Mine with 1.3 million tons. Most of this coal was shipped to electric-generating plants in the two counties, and to power stations elsewhere in Colorado.

Not all of Colorado coal producers were prosperous in the 1980s. Those that produced metallurgical coal suffered dramatic reversals. While the output of low-sulfur coal for electric utilities declined temporarily in the 1980s, the production of coking coal for the West's integrated steel plants faltered, and then ended altogether at many mines. Competition from foreign steelmakers was a key factor in bringing about the curtailment of iron and steel production in Colorado, Utah, and California. Coke-making ceased, at least in the short run, when blast furnaces closed pending adjustments to new conditions in the metallurgical trade.

Eventually, financial losses forced two steelmakers to reorganize and the third to suspend operations. The California-based Kaiser Steel Company attempted to sell its furnaces, mills, and mining properties and, when unable to do so, halted production and closed the plant in October 1982. The CF&I Steel Corporation suspended the production of pig iron in 1981. Eighteen months later, the company's management permanently shut down four blast furnaces, two basic-oxygen furnaces, and all by-product coke ovens. Thereafter, steel for the rolling mills was made from scrap metal in electric furnaces. The corporation sold its coal mines in southern Colorado, closed iron mines in Wyoming and Utah, and disposed of all properties that did not directly contribute to the new pattern of steelmaking operations.

Only the Geneva Plant at Orem survived the economic collapse of the 1980s largely intact. Faced with intense competition from abroad, plus the need to modernize the plant at an estimated cost of $1 billion, the United States Steel Corporation decided to cut back Utah operations in 1982, and three years later, to end all production. USX, as the successor to the United States Steel Corporation, sold the steel plant to a local enterprise, Basic Manufacturing and Technologies of Utah Inc. The new firm's by-product ovens, remodeled to conform to the Environmental Protection Agency's air-quality standards, were all that remained of a once-thriving coking industry in the West.

Changes in the fuel market exacted heavy social costs. Employees in the metallurgical mines of southern and western Colorado faced underemployment, at best, and in most instances the loss of jobs as mines curtailed output or shut down operations. At the same time, the rising demand from electric-generating plants led to rapid expansion in the output of boiler fuel. Mining boomtowns, particularly in Moffat and Routt Counties, experienced a heavy influx of miners and their families. Among the many consequences was a dramatic increase in social problems, including domestic disturbances, juvenile delinquency, and a variety of felonies. Sometimes, the employers assisted communities with company-financed ameliorative programs by constructing trailer parks, housing subdivisions, water and sewer lines, and recreation centers. It was important to allow residents to fill nonworking hours with constructive activities. State and federal agencies provided emergency funds to augment local resources for expanding overburdened community services.

After years of growth stimulated by Project Independence, Colorado's coal production edged downward during a period of national recession from 1982 through 1985, a reaction to turmoil in the fuel market. Another round of expansion began in the closing years of the decade. By then, because of the slower growth of electric power plants, increased output was accommodated easily by planning within the industry and the mining communities.

In Colorado, the consumption of coal moved consistently upward during the years of the energy crisis. From 1970 to 1985, it increased from 115.7 trillion to 299.1 trillion Btu, and coal's share of the energy market increased from 20.7 to 35.3 percent. Households and commercial establishments together accounted for only a small percentage of coal use. Industries claimed a larger share. Electric utilities accounted for most of the increase, as their share of the state's coal consumption jumped from 57.3 percent in 1970 to 98 percent in 1995.

The volume of solid fossil fuel devoted to the generation of electricity throughout the United States raised questions about its impact on the environment. As revealed in Figure 9, numerous environmental disturbances that adversely affected air, land, and water were associated with the production and consumption of coal. The new emphasis on coal, sanc-

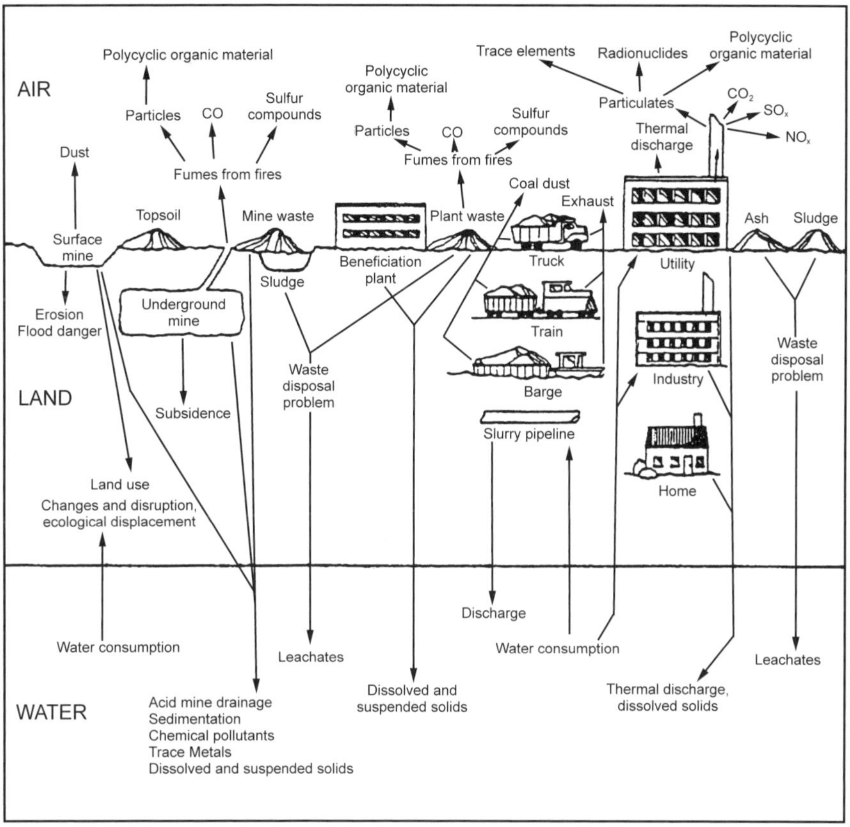

Figure 9. Environmental Disturbances From Coal-Related Activities. From U.S. Office of Technology, *The Direct Use of Coal* (1979).

tioned by Project Independence, heightened concerns, especially about deteriorating clean-air standards and the negative effects of acid rain on plant and animal life. In addition, the National Academy of Sciences warned, in 1977, of a greenhouse effect, or global warming, as the result of rising levels of carbon dioxide emissions. A major source of CO_2 in the atmosphere was the combustion of coal and oil.

Several gases that occur naturally in the atmosphere play a vital role in regulating the earth's surface temperature. They are water vapor, carbon dioxide, methane, nitrous oxide, and, in recent times, a number of man-made chemicals, of which chlorofluorocarbons are best known. These are now called greenhouse gases. They maintain the earth's surface temperature at acceptable levels by trapping the sun's heat in the atmosphere. If the natural balance of these gases is disturbed, the earth's climate is affected. If too little solar heat is retained in the atmosphere, the earth will cool. Conversely, if solar heat is trapped, unable to escape into space, the result will be global warming.

The concentration of greenhouse gases in the atmosphere had risen to dangerous levels in the past hundred years. A primary cause for that increase was thought to be the combustion of fossil fuels and the expansion of agricultural operations. By the late twentieth century, the combustion of fuels in industrial boilers, power plants, and motor vehicles was deemed responsible for about four-fifths of all greenhouse emissions in the United States. This was largely, but not exclusively, carbon dioxide. Other pollutants included sulfur dioxide and the nitrogen oxides, which were directly responsible for acid rain. Rising levels of greenhouse gases raised the likelihood of global warming.

Mounting concern about air pollution prompted a response. Congress, prodded by scientists and environmentalists, approved, and President Jimmy Carter signed into law, the Clean Air Act of 1977. It allowed only electric plants licensed or under construction prior to September 1978 to meet emission standards by burning western coal. All other enterprises had to use the best available technology, meaning electrostatic precipitators or similar devices, without regard to the sulfur content of the fuel consumed. To the disappointment of many people, the legislation of 1977 did nothing to lessen the growing volume of carbon dioxide emissions and the threat they posed in the form of global warming. In addition to air pollution, the leasing of public land for surface mining was a concern, particularly in the West. The federal government, as the largest owner of coal lands, was in a position to impose controls on the extraction of the fuel. The Coal Amendment Act of 1976, approved by Congress over President Gerald Ford's veto, discouraged the holding of federal leases for speculative purposes and mandated competitive bids. Minimum royalties of 12.5 percent were set for coal extracted from surface mines and 8 percent from underground workings.

Another concern was the potential for widespread disruption of the landscape as the result of accelerated development of western energy resources, particularly surface mining of coal and oil shale. The possibility that thousands of acres would be torn up to gain access to underlying minerals was not a happy prospect for many people. The manner in which overburden was removed and coal extracted posed a potential threat to the region's natural beauty. Widespread agitation in opposition to unregulated strip mining eventually prodded Congress into action.

The Surface Mining Control and Reclamation Act of 1977 set standards for mined lands. This law created the Office of Surface Mining to enforce uniform standards for reclamation on public and private lands. The extraction of resources had to be conducted in a manner to avoid pollution of groundwater, and operators were required to restore mined areas to approximately the original contours and to revegetate the land. A tax was levied on the production of surface mines to fund the reclamation of abandoned mining lands.

Efforts to control the negative impact of mining and consumption of fossil fuels on the environment were not always greeted with enthusiasm by mine owners and operators. Some saw the push for controls as a means of blocking economic growth. Environmental activists were often seen as a highly vocal minority of anticapitalists who opposed private enterprise in any form.

A more balanced view suggested that most environmentalists were not opposed to growth per se, but wanted to make sure that it did not inflict permanent damage on the quality of air, land, and water. In fact, there is no evidence to suggest that environmental controls of the 1970s prevented, or even perceptibly slowed, the expansion of energy production in the West. Companies that produced coal, oil, natural gas, and synthetic fuels accepted, or at least tolerated, constraints because it was good public relations, and any additional costs could be passed on to consumers.

The cost of solid fuels jumped in the 1970s, not because of environmental regulations but in response to skyrocketing oil prices. The average price of a ton of coal at the mines advanced 187 percent by the close of the decade. In the same years, the price of crude oil increased from less than $4.00 to $21.59 a barrel. President Carter's decision to decontrol oil prices in 1979, at a time when the Iranian revolution created another interruption in shipments from the Middle East, drove crude petroleum to nearly $32 a barrel in 1981. The cost of natural gas to consumers soared in a similar manner by over 600 percent per thousand cubic feet during the years 1973–1980. Higher prices stimulated domestic exploration and development of all energy resources. They also encouraged conservation.

In Colorado, the energy crisis focused attention on the need to expand the output of all fossil fuels. While the expansion of coal mining was a key aspect of Project Independence, also important were efforts to find new sources of crude oil and natural gas, and to test new technologies for the conversion of coal to gas and for the extraction of oil from shale. These activities occurred in two phases: the first as a response to the Arab embargo invoked in 1973, and the second as a reaction to still another curtailment of oil exports in 1979, when religious fundamentalists toppled Shah Mohammed Reza Pahlavi from power in Iran. Unfortunately, the emergence of new leadership in Iran coincided with a surge in oil consumption by the world's industrialized nations, making it difficult, in the short term, for energy supply to keep pace with demand. Throughout the dozen years of the crises, Colorado was an important source of oil and natural gas for domestic consumption. The production of oil slid downward, while the production of natural gas moved upward. The failure to find new, large oil fields boded ill for the state's petroleum industry.

Petroleum and gas production in Colorado followed the regional pattern. The number of drilling permits issued by the state's Oil and Gas Conservation Commission rose steadily from about 1,200 in 1974 to a high of nearly 2,400 in 1981, then fell, dipping below 1,000 in 1986. The number of active wells almost doubled from 1974 to 1980, and nearly tripled by 1985, but the production of crude oil fell by almost 21 percent from 1974 to 1980, and then stabilized in the 1980s at about 30 million barrels a year.

The accelerated pace of exploration centered in eastern Colorado. Adams, Arapahoe, Weld, Logan, and Washington Counties experienced high levels of activity, resulting in the opening of 127 oil pools. In the western portion of the state, 10 new pools were opened in Moffat and Rio Blanco Counties. Even though oil prices fell significantly commencing in 1981, another 330 pools were discovered in the 1980s. Most were in the Denver Basin, but, at the same time, exploration moved eastward, leading to numerous pools in Cheyenne, Kiowa, and Lincoln Counties.

Heightened exploration for oil and gas was largely a response to rising prices. Thanks to manipulation by the Organization of Petroleum Exporting Countries (OPEC), the price of crude oil in the United States climbed to $31.77 per barrel in 1981. At that time, some observers predicted the price could continue upward to $50.00, even $100.00, in the foreseeable future. However, rising production throughout much of the world, plus reduced demand because of conservation measures in industrialized nations, caused prices to falter and move downward. By 1985, the average domestic price of crude oil was $24.09 a barrel, a drop of about one-third in four years. This reversal undermined the joint effort by government and private industry to achieve national energy self-sufficiency.

About 40 percent of all energy consumed in Colorado during the years of the energy crisis was derived from petroleum. Consumption rose from 274.0 trillion Btu in 1970 to 330.2 trillion Btu in 1985. Oil had only a small impact on household and commercial establishments, but contributed significantly to industrial and transportation activities, accounting for more than two-thirds of the energy consumed by industries in 1975. This fell to 48.3 percent in 1985. Nearly all of the energy consumed by the state's transportation sector was petroleum in 1975 as well as in 1985, for it was the source of fuel for motorcars and other highway vehicles, diesel-electric railroad locomotives, boats, and airplanes.

Natural gas, as well as petroleum, was in short supply in the United States during most of the 1970s. Known reserves fell consistently from the mid-1960s throughout the 1970s. Unlike petroleum, the volume of imported gas was small, and did not compensate for the decline of domestic production. Natural gas from foreign sources amounted to only about 5 percent of total domestic consumption in 1977. Because of dwindling supplies, distributors frequently had to refuse service to new commercial and industrial customers and curtail, or interrupt, deliveries to established patrons.

A gas shortage occurred in many states, particularly the Northeast, in the winter of 1976–1977. This finally convinced Congress that something had to be done to increase supplies. The first response was a stopgap measure, the Emergency Natural Gas Act, which President Carter signed shortly after his inauguration in January 1977. That law empowered government agencies to reallocate gas from states where there was a surplus to others where there was a shortage.

Another winter of short supplies prodded Congress to deregulate natural gas. The Natural Gas Policy Act of October 1978 reversed a forty-year trend toward ever-increasing governmental control over prices. It distinguished between long-established and recent sources. Old gas, defined as that from wells that had been in production in 1977, was to remain regulated for a time, but at higher prices. New gas—that developed since 1977—was to be deregulated in steps over a period of seven years. By 1985, the price of all natural gas in interstate/intrastate commerce was to be determined by market forces rather than by public agencies.

Advocates of decontrol had argued for years that higher prices, reflecting competition between energy sources in the marketplace, would increase output. There was, they insisted, no lack of gas in the ground, only a lack of economic incentives to find, produce, and distribute it to consumers. By early 1979, the United States experienced a natural-gas surplus—called a "gas bubble." This happened in part because of the convergence of decreased demand caused by a switch in industrial use from gas to coal, the effects of conservation induced by higher prices, and increased output. In anticipation of higher prices, some gas previously devoted to intrastate commerce was redirected to interstate distribution.

Colorado's natural-gas output increased substantially in the years of the energy crisis. In 1974, twenty-eight counties reported production of 149.5 billion cubic feet of gas. Eleven years later, thirty-two counties produced 386.4 billion cubic feet. In 1974, Rio Blanco, La Plata, and Moffat Counties led production. By 1985, Weld County had taken the lead, followed by La Plata and Rio Blanco.

The use of gas in Colorado did not grow apace with production. Consumption fell from 275 trillion Btu in 1970 to 218.7 trillion in 1985, at a time when in-state production increased from 105.8 billion to 373.2 billion cubic feet. Residential consumption increased during the period. Dramatic shifts from natural gas to coal by industries and electric utilities, mandated by federal policies, resulted in the significant decline.

Surplus gas supplies in many parts of the country adversely affected the synthetic-fuels program. Preparations for large-scale conversion of coal into marketable gases and liquids, after experiencing notable advances, collapsed by the mid-1980s. Numerous coal-gasification projects, none of which were in Colorado, were abandoned. The one successful project, the Great Plains plant in North Dakota, had been built principally with Synfuel Corporation money. It was taken over by the Department of Energy when private sponsors withdrew.

A similar fate befell the oil-from-shale program in western Colorado. The fledgling industry went from boom to bust in less than a decade. When the Arab embargo diminished access to petroleum supplies in the Middle East, the federal government subsidized the extraction of oil from shale deposits. Fortunately, the government owned four-fifths of the oil-shale resources in the West, and was prepared to lease tracts in the Green

River formation at the time that the Arab embargo pointed to the need for alternative sources of oil.

Major petroleum enterprises acquired four tracts at a total cost of nearly $900 million, an amount that seemed outrageously high at the time to some observers. Standard Oil of Indiana and Gulf Oil Corporation leased Tract C-a, comprising 5,089 acres on the western edge of the Piceance Creek Basin in Rio Blanco County, Colorado. A partnership of Atlantic-Richfield, Ashland Oil, Shell Oil, and The Oil Shale Company (TOSCO) leased Tract C-b, 5,094 acres located thirteen miles south of Tract C-a. The bidding was more subdued for oil-shale lands in Utah. Sun Oil Company and Phillips Petroleum secured Tract U-a, and the White River Oil Shale Company (a joint endeavor of Phillips, Sun Oil, and Standard Oil of Ohio) leased Tract U-b. There were no bidders for tracts in Wyoming.

High hopes for the quick production of oil from shale on the federal leases were dashed by deficiencies in needed technology, and by uncertainties about the price of crude petroleum. Both had hindered production in previous attempts to develop the commercial potential of western oil shale. New technology required huge investments of capital, which oil corporations were willing to commit only if they could anticipate long-term high prices for petroleum.

An aboveground plant designed to produce 50,000 barrels of oil daily was estimated to cost from $5 to $6 billion, and to require up to ten years for construction. An in situ plant of the same capacity, designed to extract oil from shale underground, was estimated to cost from $3.5 to $4.5 billion. In addition, much of the requisite technology for removing kerogen from shale on a large scale and recapturing it as oil was untested, even undeveloped, at the time of the Arab boycott. There was also the need to recruit a workforce, build new communities or expand older ones, and provide schools, hospitals, sanitation, streets, police and fire protection, and commercial establishments.

Synthetic-fuels research and development gained a needed boost from the Energy Security Act, which President Carter signed into law on June 30, 1980. It created the United States Synthetic Fuel Corporation, a federal corporation with authority to encourage the production of 500,000 barrels of synfuels daily by 1987, and 2 billion barrels a day by 1992. To achieve these goals, the corporation was empowered to offer price supports, outright loans, loan guarantees, and other incentives. In addition, it had the option of asking Congress for additional money in 1990.

Table 9.1. Proposed Oil-Shale Projects in Colorado

Name of Project	*Operator*	*Method of Extraction*
Parachute Creek	Union Oil of California	Surface retort
Colony	Exxon and TOSCO	Surface retort
Paraho	Anvil Points Consortium	Surface retort
Cathedral Bluffs	Occidental Petroleum, Ashland Oil, and Tenneco Inc.	In situ
Rio Blanco	Gulf Oil and Standard Oil of Indiana	In situ and surface retort
Clear Creek	Standard Oil of California and Conic	Surface retort
Parachute Creek	Mobile Oil Co.	Surface retort
Logan Wash	Occidental Petroleum	In situ
Horse Draw	Mullet-Mineral Corp.	Surface retort
BX	Equity Oil Co.	In situ

Numerous enterprises announced intentions of investing substantial funds to plan for and, in some instances, to build plants to extract oil from shale. While a few corporations had commenced oil-shale research in the 1950s, most entered the field after the OPEC embargo of 1973. They recognized the need for expanded domestic energy supplies and hoped to capitalize on rising oil prices during what appeared to be a long-term shortage of petroleum. Most of the firms active in the shale region of western Colorado were owned by petroleum companies, either singly or in partnership with other firms.

All of Colorado's oil-shale projects were in the Piceance Creek Basin of Garfield, Mesa, and Rio Blanco Counties. Seven firms planned to utilize aboveground processing techniques in which shale had to be mined, crushed, and retorted. Each ton of shale processed in Colorado yielded about 25 gallons of oil and enormous quantities of spent material.

Retorts had many shortcomings. The cost of oil produced from shale was high because of the limited capacity of available technology. In addition, the process required the consumption of water, a scarce commodity in much of western Colorado, and the mining of shale, plus the large volume of spent material posed threats to the environment. For these reasons, some enterprises experimented with techniques for underground processing of shale.

Four firms contemplated treating shale underground. Tests indicated that in situ extraction of oil from shale was feasible. This method required that an underground deposit of shale had to be fractured and reduced to rubble before heat was applied under controlled conditions to release kerogen. That fluid was pumped to the surface, where it was stored until refined to make petroleum products.

Five of the projects had the potential of making the oil-from-shale industry a reality. Union Oil's Parachute Creek project, north of Parachute on land the company had purchased in the 1920s, planned for a total of six retorts to produce 50,000 barrels of oil a day. Commercial production was to begin in 1983.

Chevron Shale Oil Company, a joint venture of Standard Oil of California (Chevron) and Conoco, proposed to mine shale on its own land and to ship it to Salt Lake City for processing in a high-capacity retort. At an estimated cost of at least $5 billion, the Clear Creek project hoped to produce 100,000 barrels of oil a day by the mid-1990s.

The Colony project, originally a partnership of Atlantic-Richfield and TOSCO, became an Exxon venture when that firm bought Atlantic-Richfield's majority interest in 1980. On private land, the Colony project planned to spend from $2 to $3 billion to produce 500 million barrels of oil over a period of thirty years.

The Rio Blanco Oil Shale Company, a joint venture of Gulf Oil and Standard Oil of Indiana, held federal lease C-a, from which it proposed to extract oil from shale by both underground and surface methods. It was anticipated that commercial production could be achieved by 1988, with 57,000 barrels a day by in situ retorts and another 19,000 barrels a day by surface retorts. The long-term goal was 200,000 barrels of oil a day, at an estimated total cost of more than $2 billion.

The Cathedral Bluffs project, a combined effort of Occidental Petroleum, Ashland Oil, and Tenneco Inc., was located on federal lease C-b, the most northern site in the Piceance Creek Basin. Occidental Oil conducted tests with in situ production at its privately owned Logan Wash site. As many as 120 underground retorts were planned on the C-b lease over a period of thirty-five years. Full production of 90,000 barrels of oil a day was scheduled for 1991, at an estimated cost of $3 billion.

The public land leases were disappointing, for none remained active more than seven years. The Utah leases temporarily suspended operations in 1976 and shut down permanently the following year. The federal courts subsequently upheld Utah's claim to the lands and to the federal lease payments. In Colorado, on-again, off-again experiments above and below ground produced only limited amounts of oil from shale. The Rio Blanco project's experimental in situ plant, south of Rangely, produced small quantities of oil. Success in pumping kerogen from three reservoirs encouraged the partnership to initiate work on a commercial plant that produced a total of 23,000 barrels of kerogen in 1980 and 1981. However, technical problems arising from design errors slowed and eventually led to a suspension of construction. The Rio Blanco Oil Shale Company concluded that mechanical and technical problems precluded cost-effective in situ production. They hoped for better results from surface facilities for retorting shale.

Under contract with the United States Navy, the Paraho Development Company, lessee of the federal research facility at Anvil Points, mined, crushed, and retorted shale in 1977–1978, producing kerogen, which was refined by Standard Oil Company of Ohio. The fuel and lubricants were consumed in the navy's aircraft and ground vehicles. Unable to gain approval of an environmental-impact statement for the site near Rifle, Paraho relocated its operations to a site about forty miles south of Vernal, Utah. For a time it considered building another facility in Kentucky to process shale.

In the closing weeks of 1981 retrenchment became the rule in the western synfuels industry. Occidental Petroleum and Tenneco Inc., general partners in the Cathedral Bluffs project, announced the suspension of all work on their federal lease C-b until further notice. Almost immediately, Standard Oil of Indiana and Chevron (successor to Gulf Oil) issued notices of intention to indefinitely cease operations of the Rio Blanco project on federal lease Tract C-a, where operations had been stalled by uncertainties about disposing of spent shale.

The bubble finally burst when the downward slide of oil prices continued into early 1982. On May 2, a day locally as known as "Black Sunday," Exxon U.S.A., the majority partner in the Colony project on Parachute Creek, decided to abandon its commitment to the production of oil from shale. This decision resulted from an internal study that concluded the cost of the plant would be more than double original estimates, and that the cost of producing oil would be more than double the world price of oil. Exxon purchased TOSCO's minority interests, and abruptly ended operations, leaving 2,200 employees without jobs. This shook the industry to its very foundation.

Exxon's belated commitment to the Colony project in 1980 had stimulated investments and fostered exaggerated expectations for the extraction of oil from shale. The company's sudden departure accelerated widespread retrenchment as one by one the sponsors of oil-shale projects abandoned operations. Within weeks, about 10,000 people were unemployed and a mass exodus began from western Colorado. By the end of the year, only Union Oil's Parachute Creek project was alive. It remained active because of the federal government's pledge to subsidize the production of oil from shale.

In addition to oil shale, the Synthetic Fuel Corporation had encouraged the conversion of coal to gaseous and liquid fuels. Of the $200 million it awarded to applicants in 1980, one-half went to the support of enterprises that planned to convert coal into pipeline-quality gas. The three principal cooperative grants were awarded to the Great Plains Gasification Associates, in North Dakota; to Texas Eastern's project, in eastern Kentucky; and to WyCoal Gas Inc., in Wyoming. Projects under way in Colorado, Montana, New Mexico, and Wyoming hoped for subsequent federal funding.

Most of the coal-gasification enterprises did not advance beyond the planning stage. The industry was clearly in trouble by 1981, the victim of escalating construction costs, falling fuel prices, and the loss of ongoing governmental support. The price tag for production facilities nearly doubled in 1980 and 1981. At the same time, the Synthetic Fuel Corporation showed little enthusiasm for investing large amounts of public monies in a commercial synfuel industry. Funds were provided for research, but when experimental projects showed commercial promise, the agency insisted that private enterprise had to assume the burden of additional research and development. Given the rate at which fuel prices were declining in the early 1980s, that was not an acceptable avenue of economic opportunity for private corporations. With the exception of the Great Plains plant in North Dakota, funding for all coal-to-gas ventures was discontinued in the mid-1980s.

The "crash" of 1982 was a blow for Colorado's Western Slope. Before some communities had the opportunity to adjust fully to the rapid growth of the 1970s, they had to adapt to sudden, sharp contraction in the 1980s. In communities dependent on the development of synthetic fuels, unemployment escalated, causing workers and their families to leave in search of jobs elsewhere. Businesses that had expanded to accommodate the boom faced bleak futures. The few that survived had to undergo difficult adjustments. Locals relearned the forgotten lessons of their frontier past: that good times based upon extractive industries were usually short-lived. Booms were invariably followed by busts.

That was the lesson that many energy companies learned, to their misfortune. Perhaps the classic example of the energy high-flyer-crashed-to-earth was Denver's Petro-Lewis Corporation, the country's leading vendor of publicly offered oil and gas income properties. Jerome A. Lewis, a veteran Colorado oilman, founded the company in 1968 and built it into a $2 billion enterprise before plunging oil and gas prices, plus a burdensome load of debt, threatened the firm's solvency. Heavy losses forced Petro-Lewis to sell a billion dollars' worth of properties, and to reduce significantly the firm's workforce, which had peaked at 2,100 employees. In response to a lawsuit by disgruntled investors, limited partnerships were converted in 1984 into royalty trusts supported by oil and gas reserves worth an estimated $854 million. This exacerbated the firm's financial problems. Continued forced sales of properties and staggering losses pushed Petro-Lewis to the brink of bankruptcy before a buyer, a subsidiary of Freeport-McMoRan Inc., a New Orleans oil and gas producer, bought the firm and its affiliate, the American Royalty Trust, in July 1986. The Denver office closed in April 1988, at which time the last eighteen employees were terminated, all that remained of what had been one of Denver's billion-dollar oil and gas enterprises.

Denver, the acknowledged energy capital of the Rocky Mountain states, felt the full impact of the energy collapse. Colorado's principal commercial-financial center experienced painful adjustments as it responded to the changing fortunes of the fuels industry. The city had ridden an economic roller coaster to dazzling heights in the 1970s, when the metropolitan area had grown in population from 1.2 to 1.6 million. New office towers, built to house energy-related corporations, changed the skyline. The construction boom was a major contributor to prosperity. There appeared to be no limits to the potential for growth as fuel prices climbed steadily upward. It was expected that the trend would continue throughout the 1980s, and probably to the close of the twentieth century, when some industry optimists anticipated that crude oil would sell for $100 a barrel.

The city's economy plummeted into hard times in the 1980s. Most of the more than two thousand oil, gas, and synthetic-fuel enterprises cut back on operations, and a number closed their doors forever. The result was a severe recession that affected the community and the region. The collapse of the energy boom began in 1982, when prices moved downward, causing existing and planned developments to be curtailed or abandoned. Without federal funding, synthetic-fuel activities dried up and disappeared. Enthusiasm for oil and gas exploration fell off, as did output in time because of the decline of established fields, without new discoveries in similar number and size to take their place. A nationwide recession prolonged Denver's economic slump.

The end of the energy boom adversely affected petroleum, natural gas, and synthetic fuels, but coal mining remained surprisingly prosperous, given the problems that beset other energy industries. The solid fossil fuel industry owed its ongoing state of well-being to the constant, even rising, demand for coal by electric utilities throughout the western two-thirds of the nation. Because of concerns about air pollution and acid rain, economical, low-sulfur western coal was the fuel of preference at electric-generating plants during the balance of the 1980s. Coal met the challenge of the 1970s, and emerged from the energy recession of the 1980s stronger in relationship to oil and gas than it had been for several decades.

10

Beyond the Energy Crisis

The drive for national energy self-sufficiency faltered and was abandoned in the 1980s. After having supported a wide range of initiatives in the 1970s, politicians adopted the view that public programs stifled, rather than encouraged, energy production. Republican presidential candidate Ronald Reagan campaigned in 1980 on a promise to dismantle much of the federal bureaucracy, including the Department of Energy and the energy-related central planning agencies that had been created with the blessings of Presidents Richard Nixon, Gerald Ford, and Jimmy Carter.

The new administration hoped to bow out of energy programs, leaving to the private sector and a free-market economy the task of seeking a proper balance between consumption and production. After some equivocation, Reagan endorsed one exception by continuing his predecessor's policy of stockpiling oil in a Strategic Petroleum Reserve. Advocates of Reaganomics predicted that higher prices would stimulate the output of energy resources, and at the same time encourage conservation in their use. Furthermore, private programs, unlike their public counterparts, would not expand the federal bureaucracy or enlarge the national debt.

In spite of reduced government activities, there were no long-term advances in domestic oil production. Consumption of energy, particularly petroleum, after declining for a time, increased as the price of foreign oil declined. Free-market advocates should have anticipated this development, for imported oil was cheaper than the domestic energy alternative, coal. Furthermore, it was oil that fueled the transportation sector of the nation's economy, and it continued to grow at a steady rate. Petroleum and natural gas were generally the fuels of preference for all consumers, for they were more adaptable than coal to a variety of uses, and were less hazardous to the environment.

The downward trend of oil prices reflected global growth in output. Higher prices had stimulated exploration and development in Latin America, western Europe, Asia, and Africa. The resulting competition, combined with successful conservation measures in industrial nations, forced members of the Organization of Petroleum Exporting Countries (OPEC) to increase output in order to preserve their respective shares of the market. The inevitable glut undermined the world price of oil, and it remained unstable for the rest of the 1980s and well into the following decade.

Cheap oil ended the nation's search for energy self-sufficiency. The United States increased its dependence on foreign sources, in spite of risks of future interruptions of supplies from abroad. The country was no nearer to energy independence in the 1990s than it had been in 1973, when President Nixon outlined a plan to achieve that goal.

What went wrong? Why had the United States not achieved its announced goal of energy independence? The answer must be found in an examination of the energy industry, as well as in the political arena in which private-enterprise policymaking occurred.

A major obstacle to a coherent energy policy was the absence of a consensus in a country with extremely diverse energy interests. Oil and gas-producing states, in which energy producers exercised all the political influence money could buy, usually found it difficult, if not impossible, to reach an accord with coal-producing states. What benefited one group did not necessarily suit the needs of the other. States that sold fossil fuels had concerns that differed from those of states that had to buy energy. In addition, attempts to advance energy output raised the ire of highly vocal and organized groups, especially environmentalists, who were concerned that increased production and consumption of energy could threaten the quality of the nation's land, air, and water. They were determined to safeguard the remaining oil frontiers offshore along the Pacific Coast from California to Oregon and Washington, and the wilderness areas on Alaska's north shore, near the prolific Prudhoe Bay field.

The dynamics of American politics worked against the achievement of a coherent government-planned energy policy. The two political parties that shaped the actions of Congress and the White House found it impossible to agree on common actions. Partisanship, in this instance, arose largely from opposing political and economic philosophies.

Democrats, as a general rule, had few, if any, objections to central planning. They were willing to impose government actions, including controls and taxes, in the pursuit of what they identified as worthwhile national goals. In their view, it made little, if any, sense to rely on market mechanisms to ensure domestic energy supplies when the world oil market was manipulated by a handful of nations that, by means of OPEC, controlled three-fourths of the known petroleum reserves. Short-term energy security, as well as long-term national self-sufficiency, demanded special commitments through political action to gain energy independence.

Most Republicans, by contrast, rejected federal actions in favor of free-market mechanisms of the private-enterprise system. Energy was seen as a commodity, the efficient production of which did not require government regulation. Energy security could be obtained through the marketplace; private enterprise, given proper incentives, would develop ample resources to satisfy the nation's demand for energy. The appropriate role for government was to remove impediments to energy development by reducing the burden of national spending and taxes. In addition, Congress should eliminate price controls and abolish all regulations that discouraged investments in private enterprises.

An additional impediment to a policy of national energy self-sufficiency was the reluctance of the American people to make short-term sacrifices in order to secure long-term gains. Energy conservation, for example, was popular as long as oil prices spiraled upward in the 1970s, but public support largely disappeared as prices moderated and started down in the 1980s. In addition, the prospect of lower fuel costs discouraged efforts to achieve energy-efficient homes and automobiles. Manufacturers had advanced the efficiency of engines by 27 percent from 1973 to 1983, but the incentive for further gains largely vanished as oil prices fell during the balance of the decade. Posted 55-mile-per-hour highway limits were almost universally ignored, and eventually abandoned.

President Jimmy Carter had called the energy crisis the moral equivalent of war when he asked the American people to make sacrifices in order to achieve national energy self-sufficiency in ten years. His program for achieving that goal, although largely enacted by Congress, was only half-heartedly pursued, and quickly forgotten when oil, much of it imported, was readily available at reasonable prices.

In the 1980s, the bottom fell out of the petroleum market as prices plummeted by about 50 percent in the course of the decade. The upward movement of imports reflected the nation's heavy dependence on vehicles powered by internal-combustion engines for transportation. In 1985, 177.1 million vehicles, mostly automobiles, consumed 8.2 million gallons of fuel per day. This was largely gasoline. Ten years later, even with the greater fuel efficiency of passenger cars, 204.5 million vehicles consumed 9.5 million gallons of fuel per day. Jet-powered aircraft had hefty appetites for oil. As long as petroleum products were plentiful and relatively cheap, most Americans were oblivious to any potentially negative consequences of the nation's increasing reliance on foreign oil.

Another energy crisis occurred in 1990 when Iraq invaded Kuwait. The loss of 4.3 million barrels a day, sanctioned by the United Nations Security Council's embargo on oil from Iraq and Kuwait, plus concern that the conflict could spread to Saudi Arabia, drove prices upward in August and September. Higher prices provided the incentive for renewed oil exploration, particularly in the intermountain West, but when prices started downward in November, that effort was quickly abandoned.

In contrast to the oil crises of the 1970s, the Gulf War of 1990–1991 did not create significant shortages of gasoline, and American motorists did not have to wait in long lines to get to gasoline pumps. The public remained calm largely because inadequate supplies and price increases were temporary. In an effort to minimize public alarm, President George Bush authorized a test release of petroleum from the Strategic Petroleum Reserve. The Department of Energy sold 3.9 million barrels of crude oil, less than 1 percent of the reserve stored in the salt domes on the Gulf of Mexico coast. At the same time, OPEC authorized members to increase output in order to make up for the shortage caused by the loss of oil from Iraq and Kuwait. Saudi Arabia's increase alone compensated for about three-quarters of what normally would have been supplied by its northern neighbors.

In the aftermath of the Gulf War, imports continued to rise. From 1990 to 1995, they increased from 6 to 8 million barrels a day. In the latter year, nearly one-half of the petroleum consumed by Americans came from abroad. Slightly more than half the imports were from countries that belonged to OPEC. The largest suppliers were Saudi Arabia and Venezuela. Major non-OPEC sources included Canada and Mexico.

America's heavy reliance on imports should have raised concerns. It did not, at least for some time, because supplies of foreign oil were abundant, and the price was low enough to keep gasoline within the reach of most citizens. In fact, by the 1990s, falling fuel prices eroded and, in time, eliminated incentives for energy conservation. Domestic and foreign automakers had produced cars that achieved high levels of fuel economy. The Chevrolet Metro, for example, was a consistent mileage leader among fuel-efficient cars. With the price of gasoline hovering at, and even occasionally dipping below, a dollar a gallon, there was a decided shift to heavier, more powerful vehicles. Americans turned in increasing numbers to pickups, sport utility vehicles (SUVs), and minivans, most of which were less fuel-efficient than passenger cars. Light trucks, which included SUVs and many minivans, captured an ever larger share of the vehicle market.

As petroleum consumption increased, domestic crude-oil production declined. From 1985 to 1995, output in the lower forty-eight states fell by 26 percent, and in Alaska by 21 percent. In 1986, there had been 623,000 wells, each of which produced an average of 13.9 barrels of oil daily. By 1995, the number of wells had declined to 574,000, and each one produced an average of 11.3 barrels of oil daily. That year, according to the U.S. Energy Information Administration, the yield from an average domestic well, when refined, made 6 barrels of gasoline, 1.1 barrels of aviation jet fuel, 2.5 barrels of distillate fuel oil, 0.6 barrel of residual fuel oil, and 2 barrels of heavy refined products, which included lubricants, waxes, asphalt, and road oil.

In spite of the decline in domestic output, the United States was, as of 1995, the world's second-largest producer of crude oil, ranking only behind Saudi Arabia. A very large number of low-yielding wells produced in aggregate a very large volume of oil. By contrast, each of Saudi Arabia's wells had an average output of 5,800 barrels of oil a day. Petroleum was cheaper to produce abroad; hence America's increasing dependence on foreign sources of oil.

Colorado's crude-oil output, in decline since 1958, had risen briefly in the mid-1970s to 39.5 million barrels, after which there was a gradual downward trend, with no sharp contraction in production when oil prices collapsed in 1985–1986, signaling the end of the energy crisis. In the next decade, the number of active wells increased by more than 50 percent, but output hovered at about 30 million barrels a year. The yield per well fell substantially, but because of the large number, production did not dip below 29 million barrels until 1995.

Historically, Colorado's principal source of crude petroleum had been Rio Blanco County, where the Rangely-Weber pool was the leading producer. Weld County held second rank with commercial production dating from the discovery of the Greasewood pool, in 1930, but most of the output occurred after 1950. Third-ranking Washington County, active since 1952, was followed, in order, by Logan and Morgan Counties, producers since 1950. Moffat County, in sixth rank, had been a source of oil since 1924.

Until the 1980s, the principal producers of crude oil in Colorado were fully integrated enterprises. These so-called majors explored and developed new fields, produced and processed crude oil, and marketed brand-name gasoline and other products in their own automobile service stations. Since 1943, Chevron USA Inc., formerly Standard Oil of California, maintained a large presence in the prolific oil fields of Rio Blanco and Moffat Counties. Amoco (formerly Standard Oil of Indiana) had numerous wells in the Denver Basin and elsewhere in the state. Texaco Inc. was active in Rio Blanco, Moffat, and several eastern counties. Conoco Inc., the state's pioneer enterprise, was a large refiner and retailer of petroleum products.

By the time the energy crisis ended in the United States, many major companies were producing crude oil overseas rather than at home. Conoco, for example, was active as early as the 1950s in Egypt and Libya. A decade later, it was drilling wells offshore in the Gulf of Mexico, the Persian Gulf, and the North Sea. By the 1970s, Conoco had grown from a prominent regional producer and marketer of oil into an enterprise with worldwide operations.

Conoco, as a subsidiary of the DuPont corporation in the 1980s, sold many of its domestic production properties, closed two of seven refineries, and substantially curtailed retail operations in some parts of the nation. It retained, however, its Colorado refining and distribution operations, including a refinery in the north Denver metropolitan area. The chemical giant spun off its petroleum subsidiary in 1998, allowing it to function independently at home and abroad.

In Colorado, as in the United States, the primary responsibility for finding new sources of crude oil at home had passed, by the 1980s, from the majors to independent companies. The absence of large new discoveries in the lower forty-eight states since the early 1970s encouraged the majors to reduce domestic exploration and production. This opened the way for smaller firms to assume leadership in the search for oil. They were called independents because they functioned apart from the industry giants and limited activities to the production of crude petroleum.

This is not to say that all majors withdrew from Colorado. There were exceptions. Chevron USA remained the unit manager at Rangely, and Amoco Production Company played an active role in opening the Denver Basin. Independents, sometimes in partnership with integrated concerns, concentrated on finding and producing oil that was sold to other companies for refining and marketing.

Sometimes, independents were very large enterprises. The Union Pacific Railroad, after leasing for decades its immense acreage in Colorado to other firms, organized Union Pacific Resources Company, which became a leading producer of oil and gas. Commencing in 1988, it consistently ranked second in output in Colorado behind Chevron USA. In 1996, Union Pacific Resources spun off its energy subsidiary, and it was acquired, three years later, by Anadarko Petroleum Corporation, a Houston-based natural-gas exploration company.

Independents struggled to maintain solvency as oil and gas prices became very volatile in the 1990s. Low prices, particularly for gasoline, stimulated consumption of motor fuel in 1995–1996. The slump was attributed to recessions in Asia and Europe, which contributed to a worldwide glut because OPEC did not adjust its output to the reduced demand. American car owners rejoiced when prices at the pumps fell to levels, when adjusted for inflation, unknown since before the energy crises of the 1970s. Cheap gasoline wreaked havoc in the domestic oil industry. Exploration and development lagged, except offshore in the Gulf of Mexico. In order to bring costs in line with diminishing incomes, some domestic producers reduced the scope of operations and imposed widespread layoffs; others sought buyouts or merged with onetime rivals.

The resurgence of imports in the 1980s not only crippled the domestic crude-oil industry, it also dealt a severe blow to the synthetic-fuels industry. Recognizing that the energy crisis had moderated, if not ended, Congress refused to extend the life of the Synfuel Corporation in 1985. The agency had ceased operation the previous year. The few projects that had been granted loans and price guarantees were placed under the supervision of the Treasury Department. Only one, an oil-from-shale pilot plant, was in Colorado. Faced with no hope of federal subsidies, and a likely long-term decline of energy prices, other firms quietly closed their doors or turned to other activities.

Union Oil Company of California (UNOCAL) was an exception to the general trend of retrenchment in the oil-shale industry. It moved forward with operations. Construction of a pilot plant was completed in 1983, but persistent technical problems, particularly with the retort, reduced output to less than the rated capacity of 10,000 barrels per day. The Synfuels Corporation guaranteed UNOCAL a profit on the sale of every barrel from the sale of oil, and when the price fell, the company did receive, for a time, large public subsidies.

UNOCAL operated the largest and, in fact, the only commercial oil-shale project in the United States. Over a period of nearly five years, the company produced approximately 4.5 million barrels of high-quality synthetic crude oil, evidence of remarkable progress in oil-from-shale technology. In spite of generous government price supports, UNOCAL's effort resulted in substantial losses. In 1990, the production of about 1.5 million barrels of oil netted the company an after-tax loss of nearly $7 million. On March 26, 1991, faced with continuing deficits because of the low price of oil, UNOCAL shut down its Parachute Creek operation, the last of the efforts inspired by Project Independence to develop the commercial potential of the region's rich shale deposits.

The end of oil-from-shale operations raised the possibility of opening the government land in the Piceance Creek Basin for other activities. In 1997, members of Colorado's congressional delegation persuaded colleagues in both houses to transfer control of the 57,000-acre Naval Oil Reserve from the Department of Energy to the Bureau of Land Management. In that way, the property could be leased by local ranchers for livestock grazing, and to natural-gas firms for exploration.

Government-sponsored gas-from-coal programs became redundant with the growing abundance of low-priced natural gas. Only one plant emerged intact from the quest for energy independence. The Great Plains Project near Beulah, North Dakota, commenced full operation in December 1984. Using Lurgi gasifiers, the plant converted up to 14,000 tons of lignite a day into gas with an average heat content of 970 Btu per cubic foot. Other gasification plants, all experimental, were abandoned when the Synfuels Corporation failed to support their endeavors.

With the end of the energy crisis, the natural-gas industry shared domestic oil's bleak outlook. There appeared to be no relief from the low priccs, as the value of gas fell 16 percent between 1985 and 1990. This discouraged efforts to open new fields. When production finally moved upward, there was no immediate rise in prices.

The industry suffered from excessive capacity. Even as producers struggled with the problem of oversupply, federal tax laws encouraged still more production, which added to the surplus. The additional output was from compact subsurface formations that required fracturing and from coal seams.

The Crude Oil Windfall Profits Tax Act of 1980 authorized tax credits for large-scale production from unconventional sources. Credits were applicable only to wells drilled after December 31, 1979, and before January 1, 1991. Eventually Congress extended eligibility to wells drilled before December 31, 2002.

The new sources, according to the National Petroleum Council, represented 1,065 trillion cubic feet of gas reserves capable of development with existing technology. These reserves, accounting for about 20 percent of the nation's total, were located principally in the Rocky Mountain states, Alabama, and Texas. In the West, the more important source was coal-seam gas. Methane, the longtime scourge of the coal-mining industry, was accessible by vertical and horizontal wells drilled into coal measures. It could be sold to transmission companies and, in turn, to utilities for distribution to consumers as natural gas.

Colorado's natural-gas industry had shown little capacity for growth until the close of the 1980s, when coal-bed methane boosted output at a rapid rate. Production more than doubled in the years 1990 to 1995. Amoco Production Company was the first of several producers of methane from coal seams in the Ignacio-Blanco field of the San Juan Basin, and in the Raton Basin near Trinidad.

A large part of the state's natural-gas production in 1995 was coal-bed methane. A total of 962 wells in seven counties produced 240.3 billion cubic feet of methane, equal to 44 percent of all gas marketed by Colorado's producers. La Plata led with 234.4 billion cubic feet, equal to 97.5 percent of the total coal-bed methane output. Smaller amounts were produced in Garfield, Las Animas, Rio Blanco, Archuleta, Mesa, and Moffat Counties.

At the same time, twenty-one counties reported an aggregate output of 314.1 billion cubic feet of natural gas, exclusive of coal-bed methane. The leader was Weld County with 128.2 billion cubic feet, followed by Garfield with 35.5, Rio Blanco with 35.1, La Plata with 31.8, Yuma with

15.9, and Moffat with 15.8. The Wattenberg field accounted for almost 90 percent of Weld County's output.

In-state consumption of all gas, in 1995, amounted to 183.6 billion cubic feet, some of which came from adjoining states. Surplus supplies, including methane from coal seams, was exported to the Midwest and the Pacific Coast states. At this time, 92.6 percent of energy consumed by households in the United States was derived from natural gas.

Unlike oil and natural gas, coal producers escaped sharp fluctuations in output and prices with the end of the energy crisis. Coal output remained relatively stable after 1985 because of ongoing high demand for fuel by electric utilities. The construction of new power-generating facilities slowed in the 1980s. Largely for this reason, the demand for western coal grew at a modest average rate of 5 percent a year. After a brief lull, the renewed expansion of electric utilities in the 1990s further stimulated the output of the low-sulfur fuel.

In 1995, 84.6 percent of all energy consumed in the United States, measured in Btu, was derived from fossil fuels. Coal accounted for 22.1 percent, petroleum 38.1 percent, and natural gas 24.4 percent. The other 15.4 percent was contributed by nuclear power and renewable-energy sources, including hydroelectric power, geothermal energy, biofuels, solar energy, and wind energy. The industrial sector of the economy consumed 38 percent of the energy derived from fossil fuels, followed by residential and commercial (35.5 percent) and transportation (26.3 percent) sectors.

In Colorado, 94.2 percent of all energy consumed in 1995, measured in Btu, was from fossil fuels. Coal accounted for approximately 33 percent, petroleum approximately 38 percent, and natural gas approximately 29 percent. There were no nuclear power plants, and only small amounts of energy were derived from hydroelectric, biomass, and other renewable sources. The largest consumers of solid fossil fuels were thirteen electric utilities.

Colorado's utilities produced more than 35 billion kilowatt-hours of electricity in 1995. About 86 percent of that amount was attributed to coal-fired, steam-generating stations, less than 1 percent to hydroelectric facilities, and the remainder to plants fired by oil or gas. Only two generating stations—Craig and Nucla—exclusively consumed coal. The Public Service Company's Zuni plant, in Denver, burned mostly natural gas, some oil, and no coal. Other generating stations consumed coal along with some gas or fuel oil.

Approximately two-thirds of the coal consumed by Colorado power plants was produced in the state. The remainder came from surface mines in Wyoming's Powder River Valley. The largest consumer of Wyoming fuel was the Public Service Company's Comanche plant, south of Pueblo, which had been built to supply power to the CF&I Steel Corporation for making steel from scrap metal in electric furnaces.

When transportation costs permitted, Colorado, along with other western states, shipped low-sulfur coal to electric-generating plants in the Midwest and South. The demand was particularly brisk in Missouri, Illinois, Indiana, Georgia, and Texas. This market was created largely by laws designed to reduce pollution from industrial plants.

A case in point was the Clean Air Amendments of 1990. Congress stipulated that the law was to be implemented in two phases, one commencing on January 1, 1995, and the other on January 1, 2000. Phase I had as its goal a dramatic reduction in sulfur emissions (principally sulfur dioxide) from coal-burning electric power plants. They were seen as a major cause of acid rain, which was a threat to vegetation and water quality throughout much of the country. Over a period of five years the nation's 110 largest generating stations were to restrict emissions to an average of 2.5 pounds of sulfur oxide for each 1 million Btu of coal consumed. A system of allowances could be earned and sold by plants whose emissions fell below prescribed limits. Phase II limited all power plants to 1.2 pounds of sulfur emissions for each 1 million Btu of coal consumed. The legislation required, as well, the reduction of nitrogen oxide emissions by 2 million tons.

Phase I was hailed as an immediate success in reducing acid rain. Within months sulfur dioxide emissions had fallen 40 percent more than originally required. And this was accomplished at about half the anticipated cost to utilities, mainly because of the increased consumption of low-sulfur coal.

Colorado's coal production increased at a steady pace, in part because of the 1990 legislation. Twenty-three mines produced 19.2 million tons in 1990. Five years later, seventeen mines increased the total to 25.9 million tons. The number of people employed, as well as the number of mines, declined steadily, the result of new methods of extraction and the use of advanced technologies that lowered costs while producing an ever-larger volume of fuel.

Underground mines became the leading producers. The most productive units used the longwall system of mining, with extended-width panels

Table 10.1. Coal-Fired Power Plants, 1995

*No.**	*Plant*	*Utility*	*Location*	*Coal Consumed (short tons)*
1	Martin Drake	Colorado Springs	Colorado Springs	699,000
2	Ray D. Nixon	Colorado Springs	Colorado Springs	708,000
3	Arapahoe	PSC of Colorado	Denver	504,000
4	Cameo	PSC	Grand Junction	249,000
5	Cherokee	PSC	Denver	1,628,000
6	Comanche	PSC	Pueblo	2,412,000
7	Hayden	PSC	Hayden	1,727,000
8	Pawnee	PSC	Brush	2,194,000
9	Valmont	PSC	Boulder	511,000
10	Rawhide	Platte River Power Authority	Wellington	1,048,000
11	Craig	Tri-State Gen. & Trans. Assn.	Craig	4,025,000
12	Nucla	T-S G&T	Nucla	374,000
Total				16,079,000

* Identifies location of power plants on Map 5 below.

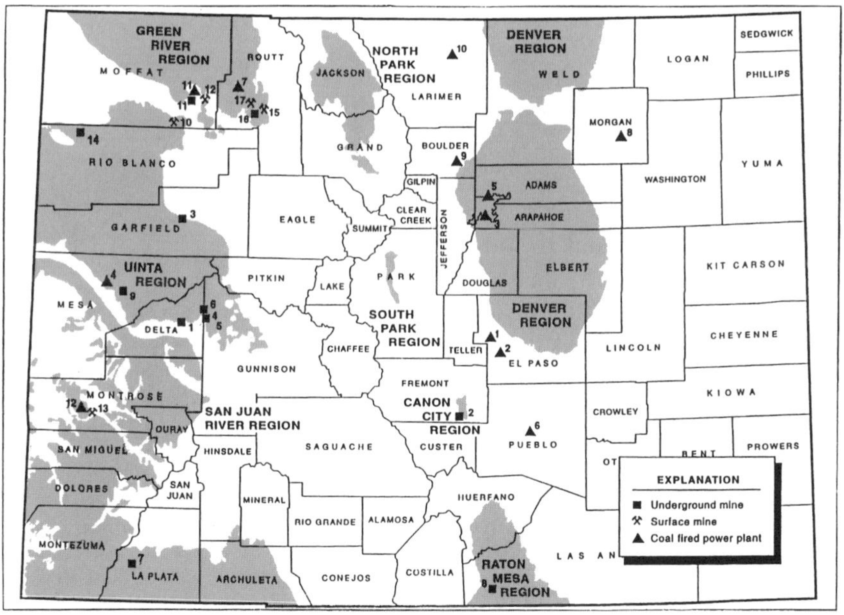

Map 5. Colorado Coal Mines and Coal-Fired Power Plants, 1995. From Colorado Geological Survey, *Colorado Mineral and Mineral Fuel Activity, 1995* (1996).

Table 10.2. Active Coal Mines, 1995

*No.**	*Name of Mine*	*Operator*	*County*	*Production (tons)*
1	Bowie No. 1 (U)	Bowie Resources, Inc	Delta	522,384
2	Southfield (U)	Energy Fuels, Inc	Fremont	302,233
3	Eastside (U)	Eastside Coal Co.	Garfield	375
4	Bear No. 3 (U)	Bear Coal Co.	Gunnison	468,002
5	West Elk (U)	Mountain Coal Co.	Gunnison	5,318,650
6	Sanborn (U)	Pacific Basin Resources	Gunnison	1,057,996
7	King Coal (U)	National King Coal	La Plata	208,642
8	Golden Eagle (U)	Basin Resources, Inc.	Las Animas	1,519,174
9	Roadside (U)	Powderhorn Coal Co.	Mesa	589,064
10	Colowyo (S)	Colowyo Coal Co.	Moffat	4,395,865
11	Eagle No. 5 (U)	Cyprus Empire Corp.	Moffat	745,074
12	Trapper (S)	Trapper Mining Co.	Moffat	2,024,186
13	New Horizon (S)	Western Fuels Assn.	Montrose	374,266
14	Deserado (U)	Western Fuels–Utah	Rio Blanco	760,255
15	Edna (S)	Pittsburg & Midway	Routt	24,088
16	Foidel Creek (U)	Twentymile Coal Co.	Routt	5,846,803
17	Seneca (S)	Peabody Western Coal	Routt	1,743,782
Total				25,900,839

* Identifies locations of mines on Map 5 above.

on coal seams up to 12 feet in thickness. As shown in Figure 10, the longwall method extracted coal with efficiency unmatched by conventional mining methods, or by continuous mining machines. In a single movement across the face of the seam, longwall shears removed coal, placing it on conveyors to exit the mine while the system advanced for the next cut. The roof was permitted to fall behind the advancing machine, causing subsidence. This precluded the use of the longwall method of extraction in populated areas. Some mines produced in excess of 5 million tons of coal a year, more than the state's entire output only a few decades earlier. Inevitably, the number of operators declined, because smaller enterprises, using older methods for extracting coal, could not complete with capital-intensive underground operations or, for that matter, surface operations.

In 1995, Colorado's production of 25.9 million tons came from twelve underground and five surface mines, as indicated below. The former contributed 69.5 percent of the output.

The largest underground mines supplied electric-generating plants. The Twentymile Mine shipped about one-quarter of its output to power

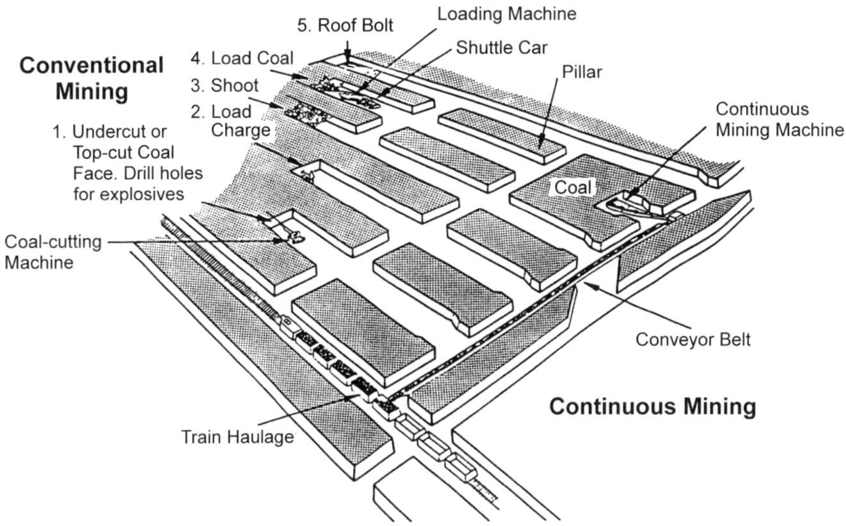

ROOM-AND-PILLAR MINING

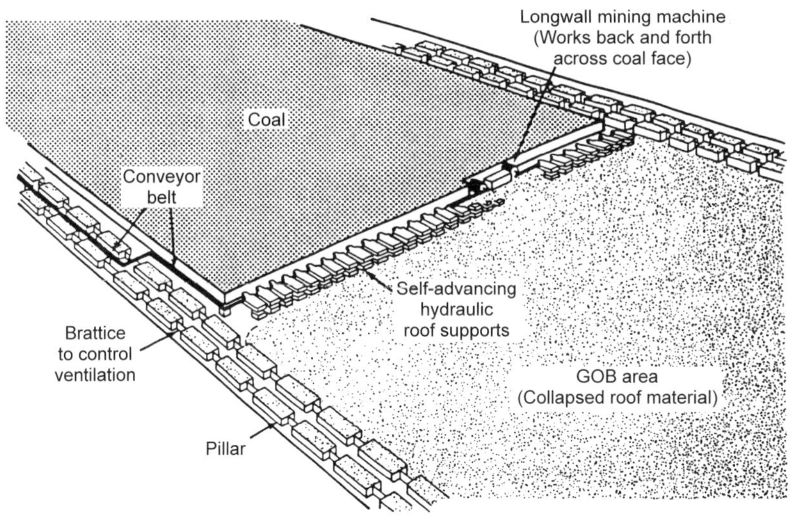

LONGWALL MINING

Figure 10. Underground Mining Systems. From U.S. Energy Information Administration, *Coal Data: A Reference* (1995).

stations in Denver and Colorado Springs, and the remainder to utilities in the Midwest and South. The West Elk Mine sent its steam and metallurgical fuel to customers in other states. The Sanborn Creek Mine shipped one-half of its production to utilities in Colorado, and the remainder to industrial customers, including utilities, in other states. Surface mining, with one exception, was confined to Routt and Moffat Counties. The New Horizon Mine, formerly the Nucla, in Montrose County, supplied the Nucla power plant. The Colowyo and Trapper in Moffat County, and the Seneca mines in Routt County, shipped to power-generating plants at Craig, Hayden, Denver, and Colorado Springs. The Colowyo Mine, south of Hayden, the state's largest surface producer, was sold in 1994 by W. R. Grace & Company to Kennecott Energy Corporation. The Edna Mine, operated by Pittsburg and Midway Coal Company for fifty years, closed in 1995 after exhausting the coal deposit.

While the output of boiler fuel for utilities increased, the production of metallurgical fuel declined in the 1980s and largely ended in the 1990s. The closing of the Kaiser plant in California, and the shift by CF&I Steel Corporation in Colorado to electric furnaces for making steel, left only the Geneva Works at Orem, Utah, as an operator of blast furnaces for making pig iron, and from it, steel. The by-product ovens were remodeled at substantial cost to conform to the Environmental Protection Agency's standards. The Utah steelworks owned its own mines in Utah and Colorado, and bought coking coal, as needed, from other producers in western Colorado. The company's Somerset Mine, in Gunnison County, remained active until 1985. After that date, the Mid-Continent Resources Corporation, operator of mines near Redstone in Pitkin County, continued shipments of bituminous coal to Wellington, Utah, where it was blended with other coals for the Geneva Works. Mid-Continent ended operations in January 1991, after financial problems prevented it from upgrading operations to conform to federal safety and environmental standards.

The CF&I Steel Corporation sold its Allen and Maxwell Mines, west of Trinidad, to the Wyoming Fuel Company, a wholly owned subsidiary of K-N Energy Inc. That firm changed the name of the properties to Golden Eagle, and hoped to sell its metallurgical coal to industrial consumers. Entech Inc., a subsidiary of Montana Power Company, purchased the property in 1991. Entech's successor, Basin Resources Inc., entered into a contract to supply the Tampa Electric Company with up to 1.2 million tons of coal a year. The mine's output rose to 1.8 million tons in 1995, but a bad roof and

Longwall mining equipment in operation at the Allen Mine, 1971. Hydraulic jacks supported the roof above the coal, while an operator used a mechanical shear to strip coal from a vein several hundred yards in width. The coal was mechanically loaded into cars, or onto conveyors, to be moved from the mine. The shears advanced for another cut as the jacks allowed the roof to subside behind the operator. Courtesy, Bessemer Historical Society, Pueblo.

geological faults dramatically increased operating costs. After expenses exceeded income by $9.5 million in the first nine months of 1995, the company halted operations, ending, after more than a century, the production of coal in southern Colorado. This was a severe economic blow to the area, for the Golden Eagle had been the largest employer in Las Animas County.

In the absence of a metallurgical industry, the only large consumers of coal in Colorado were electric utilities. Rising demand by power plants propelled the state's coal production upward to new records year after year. The largest non-utility users were manufacturers of cement and paving materials. Households and commercial customers, once dependent on the output of the state's mines, consumed less than 1 percent.

Colorado's energy industry experienced troubled times after 1995. Coal mining appeared to be more stable, but not necessarily more prosperous, than petroleum and natural gas. A handful of mines dominated the industry. Their numbers dropped from seventeen in 1995 to twelve in 1999, as output increased from 25.9 to 29.98 million tons. By the latter year, almost half of the output was consumed in Colorado, mainly by power plants, and most of the remainder was shipped to power plants in twenty-two states. A small amount was shipped to foreign countries, mainly Mexico, Japan, and Taiwan.

High-tech mines, whether on the surface or underground, were the big producers. Kennecott's Colowyo Mine, in Moffat County, produced 5.9 million tons in 1999, using monstrous machines to extract coal from shallow seams. A dragline with a 60-cubic-yard bucket removed the overburden. Thirty-five-cubic-yard loaders filled 240-ton trucks that transported coal to a preparation plant for cleaning and blending, prior to shipment by unit trains to electric power stations. In 1999, the state's largest mine extracted coal from underground seams. RAG American Coal's Twentymile Mine (Foidel Creek) employed longwall machinery to produce 8.6 million tons.

For a time, prosperity eluded Colorado's petroleum industry. The output of crude oil fell by 10 percent in 1996, another 7 percent in 1997, and again by almost 10 percent in 1998, declining that year to 22.1 million barrels. In 1998, thirty-one counties reported commercial production. Rio Blanco County was the leader, as it had been for more than forty years, followed, in order, by Weld and Cheyenne Counties. Those three counties accounted for 86.4 percent of the state's output.

Dwindling production was accompanied by lower prices. The average price for oil at the wellhead in 1998 fell to less than $10 a barrel, down dramatically since 1991. This meant lower prices for gasoline. In the Denver metro area, self-service stations sold unleaded regular at less than a dollar a gallon. Because of low prices, many producers shut in wells, contributing further to the decline in output. At the same time, exploration and development fell at a steady rate. The industry retrenched, with widespread layoffs, and weaker companies sought mergers with stronger enterprises in an effort to stave off collapse.

An abrupt reversal occurred in 1999. In January the average price of oil in the global market was $11.11 a barrel; one year later it had almost tripled. The rapid advance was attributed principally to OPEC's adoption of production limits in an effort to adjust output of crude to slumping demand. At the same time, global prosperity revived as recessions ended in

Asia and Europe. Rising demand and falling supply caused oil prices to advance to about $30 a barrel, and to remain at or near that level throughout 2000. At the year's end, shrinking crude oil inventories suggested that higher prices were likely in the new century.

Following the example of oil, natural-gas prices moved downward until the mid-1990s. In January 1995, the wellhead price of gas fell to a three-year low. All producers suffered, but the slump was particularly hard on enterprises located in the Denver Basin, where aging wells had small profit margins. Ongoing low prices discouraged exploration and development.

Financially stressed corporations turned to consolidations and mergers to strengthen their positions in an unstable market. This goal was pursued by large major as well as smaller independent companies. Amoco Production Company, the nation's sixth-largest producer of crude oil and a leader in natural gas, merged with British Petroleum PLC. British Petroleum–Amoco PLC remained the principal operator in the Ignacio-Blanco field of La Plata County. Amoco had previously sold its holdings in the Denver and Raton Basins, and its Denver office for overseeing regional exploration and development had been combined with similar operations in Houston. BP-Amoco's subsequent attempt to acquire Atlantic-Richfield Company prompted opposition from the Federal Trade Commission. Approval was contingent on BP-Amoco's sale of Arco's holdings on Alaska's North Slope.

Another merger of national importance, one that had little direct impact on Colorado, was that of Exxon Corporation, the country's largest petroleum enterprise, and Mobil Corporation. After some equivocation, the Federal Trade Commission approved the merger in November 1999 on condition that two enterprises sell about 2,400 retail units, plus other facilities, in eastern states. This and other mergers raised widespread fears that the industry giants could, because of the size and scope of consolidated operations, reduce competition and arbitrarily raise the cost of petroleum products in the United States.

Colorado's natural-gas industry quickly rebounded from the slump of 1995. Production increased steadily from 556 billion cubic feet in 1995 to 692 billion cubic feet in 1998. In the same period, the average wellhead price of gas increased from $1.19 per thousand cubic feet in 1995 to $2.29 per thousand cubic feet in 1997, before falling to $1.95 in 1998. Coalbed methane was the key to the industry's growth. Its share of total gas production increased from 39 percent in 1995 to 54.5 percent in 1998.

The Ignacio-Blanco field was the focal point of activity, followed by Las Animas County.

In January 1999, the Colorado industry was pessimistic about the long-term outlook for gas. However, prices, driven by sharply rising demand and nonelastic supply, moved upward and doubled by mid-2000. Six months later, the twentieth century ended with the American people caught up in another energy crisis as the price of scarce gasoline, fuel oil, and particularly natural gas soared to unprecedented heights.

The new energy crisis had some of the features of the previous one. The demand for energy had grown in the 1990s, as in the 1970s, faster than supply. The natural-gas price spike in 2000, according to industry spokesmen, resulted from the failure of supply to keep pace with growing demand. Persistent low prices earlier in the decade had discouraged exploration for and development of new wells. Under more favorable circumstances, new wells would have provided gas to keep pace with an expanding market.

There were, in fact, other factors that contributed to the dramatic rise in gas rises in the closing weeks of 2000. For some time, regulators, gas producers, and environmentalists had pressured electric utilities to switch from coal to natural gas. In order to avoid costly lawsuits for alleged violations of clean air standards, some older plants shut down or converted to clean-burning gas, and new generating capacity, almost without exception, consumed gas. Utilities, by turning to gas, consumed a large amount of fuel at a time when production had fallen to dangerous levels. After three relatively warm winters, colder temperatures in the final two months of 2000 put enormous strains on available gas supplies.

The imbalance between supply and demand was especially acute in the Denver metropolitan area because Excel Energy Company, formerly Public Service Company of Colorado, lost its only facility for storing gas to meet abnormal demand during the colder months of the year. The Leyden Mine, in Jefferson County northwest of Denver, had stored up to 3 billion cubic feet of gas since the 1960s. It gave the vendor the opportunity to buy during summer months, when prices were moderate, for use in the winter, when the gas prices, reflecting increased demand, rose to higher levels.

The decision to close the facility was in response to complaints about leaking gas from homeowners in the vicinity of Leyden, and from Jefferson County authorities who questioned the safety of the facility. In December

2000, the Colorado Public Utilities Commission, at the request of Excel Energy Corporation, approved a plan for ending storage in the mine. Closure was to be completed in 2001, or shortly thereafter, leaving the supplier without the means to store natural gas to meet abnormal demand for fuel in the region. The supply issue may be further complicated by the fact that Excel Energy Corporation has plans in place to build additional gas-fired, electric-generating plants in the Denver metropolitan area.

The erection of natural gas–fired electric plants reflects growing public concern about the negative impact of coal and oil combustion on the environment. This has raised questions about the future of those fossil fuels. Coal-fired, steam-generating electric plants in the United States are responsible for about two-thirds of the sulfur dioxide and a large part of the carbon dioxide annually released into the atmosphere. In addition, coal, as well as oil, contributes significant amounts of other pollutants. Motor vehicles powered by gasoline and diesel fuel are the second-largest source of air pollution. They have traditionally caused brown clouds and smog that frequently obscure urban skylines from November through March.

In spite of rising concern, Colorado's heavy reliance on coal and oil will probably continue for some time. How far into the next millennium will depend on several factors that cannot be fully anticipated at this time. One is the availability of natural gas at a reasonable price as a substitute for coal to generate electricity. Another is potential technological advances in the preparation and use of coal and oil to reduce or eliminate pollutants in emissions from combustion. A third is the development of nonpolluting alternatives for coal and oil.

The future of solid fossil fuel is a frequent topic of debate. Environmentalists want clean-fuel technologies designed to reduce costs and lower the volume of pollutants released during combustion. In the pursuit of this goal, the coal industry, in the short run, may lose ground to natural gas. The Energy Information Administration projects a 50 percent rise in the use of natural gas by 2020. The reason for this is simply that natural gas generates far less pollution than coal, or for that matter, oil. The amount of carbon dioxide released by combustion, in relation to the heat produced, ranks coal and gas, respectively, as the highest and lowest polluters.

The Energy Information Administration suggests that the long-term demand for solid fossil fuel may be dictated more by economics than by environmental concerns. Natural gas will increase in cost as supplies inevitably become scarce. National coal production, 1 billion tons in 1990, is expected to increase to 1.5 billion tons in 2010. More than 80 percent of the increase will be consumed in the United States, principally by electric utilities; the rest will be exported. Four-fifths of the new output will be low-sulfur coal from the intermountain states.

In light of the energy crisis in the winter of 2000–2001, the case for solid fossil fuel may be made in the short, rather than the long, run. In California, the cost of gas-generated electricity suddenly doubled, even tripled and quadruped, threatening to put the price of electricity beyond the means of a large segment of the population. This condition may be only temporary, but it does highlight concerns about the future of natural gas as opposed to coal for generating electric power.

If solid fuel retains its dominant role nationwide in the generation of electric power, it will have to clean up its act by significantly reducing the pollution now associated with combustion. While many environmentalists may see the idea of clean coal as an oxymoron, people in the mining industry believe that the combustion of coal does not have to be harmful to the environment. This has been the goal of the U.S. Department of Energy's Clean Coal Technology Demonstration Program, by means of which, since 1986, public and private funds have supported developmental projects for advanced coal-burning technologies. The Energy Policy Act of 1992 authorized a continuation of the federal government's contributions beyond the original termination date of 1994 in order to keep alive the goal of nonpolluting coal.

A solution to the problem of pollution from motor vehicle emissions has been, and will continue to be, elusive. By the 1990s, concern for pollution caused by gasoline- and diesel-powered engines focused attention on alternative (nonpetroleum) fuels for transportation, such as electricity, ethanol, methanol, natural gas (methane), and propane. With the exception of propane, which is extracted mostly from petroleum rather than natural gas, they will compete with one another and with emerging new technologies—for example, gasoline-electric and diesel-electric power systems for motor vehicles.

This does not mean that petroleum-based fuels will be abandoned. Quite the contrary. Recognizing that petroleum is, and for some time will be, the principal fuel for motor vehicles, the Clean Air Act of 1990 instructed the EPA to take action to reduce ozone and toxic gases generated by automobile exhausts. To this end, the EPA has worked successfully with the petroleum industry to advance the cause of clean fuels. To date, oxygenates and reformulated gasoline, according to government agencies, have reduced pollution in several urban areas.

Concern about pollution has stimulated interest in renewable sources of energy. Experiments with wind, solar, geothermal, and hydroelectric power (hydropower), as well as energy derived from biomass and municipal solid wastes, have given reason for optimism about future large-scale use. They are widely viewed as nonpolluting, but only wind, solar, geothermal, and hydropower actually fit that description. Hydropower, which accounted for 86 percent of electric generation by renewable sources of energy in 1990, is expected to grow only slightly in the next decade.

Electric-generating capacity in the United States was 732 gigawatts in 1990, and is expected to increase to as much as 930 gigawatts by 2010. Incentives written into the Energy Policy Act of 1992 will ensure growth in the generating capacity using wind, solar, and geothermal power, and their collective contribution to the nation's total will, it is estimated, increase from 12.3 gigawatts in 1990 to 33.8 gigawatts by 2010. It is anticipated that energy from biomass and municipal solid waste will experience rapid growth.

Nuclear power accounted for 22 percent of the electricity generated in the United States in 1995. Only 3 of the 109 active units were in the intermountain West, all in Arizona. The Public Service Company of Colorado constructed a nuclear plant north of Denver in the 1980s, but technical problems and large cost overruns prompted conversion to natural gas. There have been no orders for nuclear plants since 1978, but new designs incorporate technological advances and enhanced safety features for future development. The Energy Policy Act of 1992 revised licensing procedures in an effort to encourage construction of more nuclear generating capacity. As the demand for electricity grows, nuclear plants on which construction has been deferred in the past may be completed, and new plants may be authorized.

According to the Energy Information Administration, only about 6 percent of the energy consumed in Colorado in 1997 was credited to renewable resources. Hydropower and biomass accounted for nearly all of that amount. Consumption was insignificant for energy produced by wind, geothermal, solar, and nuclear sources. However, renewable energy and nuclear power may be the wave of the future, especially if the United States makes a solid commitment to the reduction of greenhouse-gas emissions.

This will happen if the Kyoto Protocol of 1997 is fully implemented. That document calls for thirty-nine industrialized nations to reduce emissions of greenhouse gases to an average of about 5 percent below 1990 levels, with compliance scheduled for the years between 2008 and 2012.

Because 83 percent of greenhouse-gas emissions in the United States are produced by the combustion of fossil fuels, the Kyoto Protocol, if fully implemented, will have far-ranging effects on energy markets. Reductions in greenhouse gas may be achieved by lowering the demand for energy, by adopting energy-efficient automobiles and other machines, and by switching to less carbon-intensive or noncarbon fuels. Any or all of these actions will require a dramatic curtailment in the consumption of fossil fuels. This could lead to higher energy costs, lower industrial output, and widespread unemployment, a recipe for economic instability.

Thirty countries have ratified the protocol. However, it appears unlikely at this time that the United States will do so. The Sixth Conference of the Parties to United Nations Framework Convention on Climate Change (COP6), meeting at The Hague in the Netherlands in November 2000, failed to win agreement on a plan for implementation. Since the U.S. Senate had overwhelmingly indicated dissatisfaction with the Kyoto Protocol, it was not surprising that the country's representatives refused to endorse targeted reductions within the stipulated time frame. Shortly after taking office in 2001, President George W. Bush refused to commit the United States to the Kyoto Protocol for the foreseeable future.

Mounting evidence of global warming suggests that coordinated worldwide efforts are needed to limit pollution. The time may be fast approaching when the American people must weigh the projected dislocation of dramatic change against the dangers of global warming and rising health risks attributed to unrestrained consumption of fossil fuels.

Bibliographical Essay

High Altitude Energy is a history of the fossil-fuels industries and some aspects of their impact on the people of Colorado. In order to serve the needs of a society, coal, oil, and gas had to be mined and marketed before they were converted into energy, the consumption of which contributed to a high standard of living characterized by rising levels of comfort and convenience. This study is, at best, an introduction to an enormously complex subject. It is hoped that it will stimulate additional investigations. In an effort to encourage further research, the following bibliographical essay identifies the principal sources the author used in the preparation of this work.

Background

It is important to remember that fossil-fuels industries in Colorado, and the consumption of energy derived from them, did not occur in a vacuum. The national and international context is the focus of Sam H. Schurr and Bruce C. Netschert's *Energy in the American Economy, 1850–1975* (Baltimore: Johns Hopkins University Press, 1960), with attention to how and why energy production and consumption increased over a period of a century. The policy context for accelerating fossil-fuels production and consumption is the subject of John G. Clark's *Energy and the Federal Government:*

Fossil Fuel Policies, 1900–1946 (Urbana: University of Illinois Press, 1974) and Richard H. K. Vietor's *Energy Policy in America Since 1945: A Study in Business-Government Relations* (Cambridge: Cambridge University Press, 1984).

Hubert E. Risser's *The Economics of the Coal Industry* (Westport, CT: Greenwood Press, 1958, 1976) looks at how the bituminous coal industry is basic to the modern industrial economy. Other important aspects of coal mining are treated by C. L. Christenson's *Economic Redevelopment in Bituminous Coal: The Special Case of Technological Advance in United States Coal Mines, 1930–1960* (Cambridge, MA: Harvard University Press, 1962) and James P. Johnson's *The Politics of Soft Coal: The Bituminous Coal Industry from World War I Through the New Deal* (Champaign: University of Illinois Press, 1979).

A large body of literature is devoted to the origin and evolution of the oil industry in the United States. Particularly informative are Leonard M. Fanning's *The Rise of the American Oil Industry* (rev. ed., New York: Harper and Brothers, 1948); Gerald Nash's *United States Oil Policy, 1890–1964: Business and Government in the Twentieth Century* (Pittsburgh: University of Pittsburgh Press, 1968); Harold F. Williamson et al., *The American Petroleum Industry, 1859–1959* (2 vols., Evanston, IL: Northwestern University Press, 1959–1963); Edgar Wesley Owen's *Trek of the Oil Finders: A History of Exploration for Petroleum* (Tulsa: American Association of Petroleum Geologists, 1975); and Bill D. Berger and Kenneth E. Anderson's *Modern Petroleum: A Basic Primer of the Industry* (Tulsa: Oil & Gas Journal, 1978).

The origin and evolution of the natural-gas industry, which gained national stature long before it emerged as an industry in Colorado, is treated in numerous books. Particularly useful are J. C. Youngberg's *Natural Gas: America's Fastest Growing Industry* (San Francisco: Schwabacher-Frey Company, 1930); Alfred M. Leetson, John A. Crichton, and John C. Jacobs's *The Dynamic Natural Gas Industry* (Norman: University of Oklahoma Press, 1955); and Arlon Tussing and Bob Tippee, eds., *The Natural Gas Industry: Evolution, Structure, and Economics* (2d ed., Tulsa: Pennwell Books, 1995).

Archives and Manuscript Collections

Archival and manuscript collections are important sources for tracking the evolution of the fossil-fuels industries in Colorado in the nineteenth and twentieth centuries. The business archives of the CF&I Steel Corporation, including the records of its predecessor enterprises, are in the possession of the successor firm, the Rocky Mountain Steel Mills. Current plans call for turning these materials over to the Bessemer Historical Society, and they will be housed at or near the steel plant in Pueblo. CF&I was the largest producer, distributor, and consumer of coal and coke from the 1870s until about 1960.

Particularly valuable are two in-house histories of the firm's operations. "Predecessor Companies of the Colorado Coal [Fuel] & Iron Co.—The Colorado Improvement Co.; The Southern Colorado Coal & Iron Co.; The Colorado Steel Works Co.; and The Colorado Coal and Iron Company," a 214-page typescript, prepared in 1923, contains statistical and other data taken from financial, production, and sales reports. A five-page typescript, "Statement of coal mined by predecessor companies prior to November 1, 1892, the Colorado Fuel and Iron Company from November 1, 1892 to June 30, 1936, and The Colorado Fuel and Iron Corporation from July 1, 1936 to December 31, 1965," is a brief summary of fuel operations.

Two collections in the Colorado Historical Society, Denver, supplement the Bessemer Historical Society's CF&I business archives. The Colorado Fuel and Iron Collection contains letters and documents principally for the years from 1872 to 1903. The bulk of the materials are correspondence, annual reports, and minute books of affiliated and subsidiary enterprises. The Jesse Floyd Welborn Collection contains copies of letters, reports, scrapbooks of clippings, and other materials for the years from 1901 to 1936. Welborn, president of the Colorado Fuel and Iron Company from 1909 to 1929, was the spokesman for the state's coal-mining industry during the strife-torn miners' strike of 1913–1914. His communications articulate the coal industry's perception of the labor dispute and its aftermath.

The Colorado Historical Society's Warwick M. Downing Collection offers valuable insights into Colorado's petroleum industry. Downing (1881–1964) was a mining attorney, independent oil operator, and conservationist. A prominent public servant, he served as chairman of the Colorado Oil and Gas Conservation Commission from 1951 to 1963. He was a member of the Interstate Oil Compact Commission, the National Petroleum Council, and the Rocky Mountain Oil and Gas Association. The collection contains correspondence, reports, clippings, and other materials related principally to the oil and oil-from-shale industry in the West.

The Western Historical Collections, Norlin Library, the University of Colorado at Boulder, houses collections that relate to coal mining. They include papers of the Coal Project, the Western Federation of Miners, and the Rocky Mountain Fuel Company. The latter collection, for the years from about 1905 to 1930, includes maps and reports. The firm's mines were located throughout the state, with a heavy concentration in Boulder and Weld Counties.

The Petroleum History Collection in the American Heritage Center, University of Wyoming, Laramie, is a vital source of data about the early development of the oil industry in Colorado. The papers of A. M. Cassiday reveal the origin and early evolution of the oil production and distribution from the Cañon City–Florence area, what became the Fremont County field. Other collections contain materials that highlight the evolution of the Continental Oil Company in Colorado and its role as an instrument of the Standard Oil monopoly in distributing the petroleum products of refineries at Florence throughout the Rocky Mountain region. The center houses the Warwick M. Downing Collection, which contains mainly duplicates of the collection of the same name in the Colorado Historical Society, Denver.

Monographs and Articles

Numerous books are devoted, at least in part, to the history of Colorado's fossil fuels. By far the largest number pertain to coal mining. Others treat the origin and evolution of the petroleum, natural-gas, and oil-from-shale industries.

Two books that stress one corporation's dominance of the regional fuel trade are H. Lee Scamehorn's *Pioneer Steelmaker in the Western: The Colorado Fuel and Iron Company, 1872–1903* (Boulder, CO: Pruett Publishing Company, 1976), and *Mill and Mine: The CF&I in the Twentieth Century* (Lincoln: University of Nebraska Press, 1992). Other books about CF&I invariably dwell on labor relations, particularly the events of 1913–1914 culminating in what the United Mine Workers of America called the "Ludlow Massacre." The first work to focus on the violent strike was Barron B. Beshoar's *Out of the Depths: The Story of John R. Lawson, A Labor Leader* (Denver: The Colorado Labor Historical Committee of the Denver Trades and Labor Assembly, 1943); probably the best is George S. McGovern and Leonard F. Guttridge's *The Great Coalfield War* (Boston: Houghton Mifflin Company, 1972).

In *From Trappers to Tourists: Fremont County, 1830–1950* (Palmer Lake, CO: The Filter Press, 1972), Rosemae Wells Campbell looks at the coal industry in what was called the Cañon City district. It was the source of fuel that was widely consumed by Colorado households, as well as by businesses and factories. The mining camps of the same district are the subject of Antoinette V. Cresto's *King Coal: Mining in Fremont County* (Florence, CO: Florence Citizen, 1980). In *The Elusive Dream: The Relentless Quest for Coal in Colorado* (Glenwood Springs, CO: Gran Farnum Printing and Publishing Company, 1983), Anna Johnson and Kathleen Yajko examine the industry and the mining camps of Garfield and Pitkin Counties. Duane A. Smith's *When Coal Was King: A History of Crested Butte Colorado, 1880–1952* (Golden: Colorado School of Mines Press, 1984) looks at the high-grade bituminous and anthracite mines and ovens at what is now a well-known ski resort in Gunnison County.

Three important studies examine coal mining in the southern field. William H. McKenzie's *Mountain to Mill: The Colorado and Wyoming Railway* (Colorado Springs, CO: MAC Publishing, 1982) looks at the Colorado Fuel and Iron Company's wholly owned railroad, which served the steel plant at Pueblo, iron mines in Wyoming, and coal mines in Colorado. A strong focus of the book is the transportation services that were vital to the success of coal mines of Huerfano and Las Animas Counties, and beehive coke oven plants in the Purgatoire River Valley near Trinidad. In *Life and Death of a Coal Company and the Story of One Man's Battle for Its Life* (Denver: Ron's Graphic Service, 1987), Dolores Plested, a veteran journalist, examines the Bear Canyon Coal Company and the operations of its mine west of Ludlow. Rick J. Clyne's *Coal People: Life in Southern Colorado's Company Towns, 1890–1930* (Denver: Colorado Historical Society, 1999) stresses the social history of coal mining in Huerfano and Las Animas Counties.

A comprehensive social history of coal mining is Eric Margolis's "Western Coal Mining as a Way of Life: An Oral History of the Colorado Coal Miners to 1914," *Journal of the West* 24 (July 1985): 1–128. This study was an outgrowth of the Coal Project, a concerted effort to collect oral history interviews, photographs, and other historical data by the Institute of Behavioral Science at the University of Colorado, Boulder. Another groundbreaking book was James Whiteside's *Regulating Danger: The Struggle for Mine Safety in the Rocky Mountain Coal Industry* (Lincoln: University of Nebraska Press, 1990).

The publication of Phyllis Smith's *Once a Coal Miner: The Story of Colorado's Northern Coal Field* (Boulder, CO: Pruett Publishing Company, 1989) turned the spotlight of history on the state's long-neglected field in Boulder and Weld Counties. In a similar fashion, Duane A. Smith's *Rocky Mountain Boom Town: A History of Durango* (Albuquerque: University of

New Mexico Press, 1980) revealed the contributions of locally produced coal and coke to the economic growth of Durango and southwestern Colorado. The impact of coal on the Durango area is the subject of Norman J. Bender's "The Coal Mining Industry and Related Coal Mining Camps in La Plata County, Colorado" (unpublished paper, 1967, 62p.), a copy of which is in the author's possession.

There are no book-length studies of coke manufacturing in Colorado. Scamehorn's two volumes on the CF&I Steel Company and its predecessor enterprises discusses the origin and evolution of the firm's beehive oven plants in Las Animas, Gunnison, Garfield, and Pitkin Counties, and the by-product coke ovens later erected at the steel plant in Pueblo. The CF&I's *Camp and Plant*, issued from 1901 to 1904, and a successor magazine, *The Colorado Fuel & Iron Company Industrial Bulletin*, published from 1915 to 1929, carried articles on the company's mining and coking plants. The origin of Cokedale, seven miles west of Trinidad, one of the state's few extant coal and coke camps, is the subject of W. B. Lloyd's "The Cokedale Mine," in Colorado State Inspector of Mines, *Biennial Report, 1907–1908* (1909): 152–155, and Kenneth Guiterman's "Mining Coal in Southern Colorado: The Carbon Coal and Coke Company Is Operating a Model Plant at Cokedale," *Engineering and Mining Journal* 88 (November 20, 1909): 1009–1015.

There are few book-length studies about the petroleum and natural gas in Colorado. Fred S. Jensen, Henry H. R. Sharkey, and Daniel S. Turner, eds., describe the mid-twentieth-century status of the industry in *The Oil and Gas Fields of Colorado: A Symposium* (Denver: Rocky Mountain Association of Geologists, 1954). Russ Roundtree's *Western Oil Reporter's Rocky Mountain Oil History* (Denver: Hart Publications Inc., 1984) examines the origin and evolution of the oil industry in Colorado, as well as other states in the Rocky Mountain region. The oil industry at Florence was an important part of Campbell's *From Trappers to Tourists: Fremont County, 1830–1950*, and H. Lee Scamehorn's "In the Shadow of Cripple Creek: Florence from 1865 to 1910," *Colorado Magazine* 55 (spring-summer 1978): 205–229. Duane Smith's *Colorado Mining: A Photographic History* (Albuquerque: University of New Mexico Press, 1977) has a heavy emphasis on hardrock mining, but looks also at coal, oil, gas, and oil shale.

The Continental Oil Company, Colorado's pioneer firm, has sponsored two histories of its corporate operations. *Conoco: The First One Hundred Years* (New York: Dell Publishing Co. Inc., 1975) and Russ Banham's

CONOCO: 125 Years of Energy (Lyme, CT: Greenwich Publishing Group Inc., 2000) follow the company's evolution from a regional distributor of oil to an international corporation engaged in producing, refining, and marketing petroleum products. Two videos—the Rocky Mountain Association of Geologists' "Lighting the Frontier: The Story of Colorado's Florence Oil Field" (1995) and Conoco's "125 Years of Energy: The Conoco Story" (2000)—offer valuable insights into the origin and evolution of the petroleum industry in Colorado.

Three studies examine the manufactured gas industry in Colorado. C. L. King's *History of the Government of Denver, with Special Reference to Its Relations with Public Service Corporations* (Denver: Fisher Book Company, 1911) looks at, among other things, the city's response to proliferating local gas companies in the late nineteenth century. David J. McCanne's "Recollections of a Civil Engineer," *Colorado Magazine* 23 (May 1946): 134–136 and (July 1946): 171–178, recounts the construction and operation of gas retorts in Gunnison. Ellen Kingman Fisher's "One Hundred Years of Energy: Public Service Company of Colorado and Its Predecessors, 1869–1969" (Ph.D. dissertation, University of Colorado, Boulder, 1986) was the source for her "Power's Dynamo Unloosed: Henry C. Daugherty and the Denver Gas and Electric Company," published in *Essays and Monographs in Colorado History* 14 (1994): 76–96. A short history of a pioneer producer and distributor of natural gas in northwest Colorado is W. T. Nightengale's *The Mountain Fuel Story: Growth of a Natural Gas Company* (New York: Newcomen Society of North America, 1963).

Booms and busts are common themes in studies of oil-from-shale programs in western Colorado. William E. Beilke's "Colorado's First Oil Shale Rush, 1910–1930" (Ph.D. dissertation, University of Colorado–Boulder, 1983) is an in-depth study of the initial boom, concentrating on how it occurred and why it failed to create a commercially viable industry. Paul L. Russell's *History of Western Oil Shale* (East Brunswick, NJ: Center for Professional Advancement, 1980) looks at attempts to create a commercial shale industry during World War I and the 1920s, and again during and after World War II. Andrew Gulliford's *Boomtown Blues: Colorado Oil Shale, 1885–1985* (Boulder: University Press of Colorado, 1989) covers the full sweep of oil-from-shale endeavors, but excels in its treatment of the shale boom of the 1970s and the collapse of the 1980s.

The energy crisis of the 1970s generated a very large body of literature devoted to pointing fingers of blame at parties allegedly responsible for the shortage of petroleum in the United States. More responsible authors asked why demand outran supplies of coal, petroleum, natural gas, and electricity long before the Organization of Petroleum Exporting Countries (OPEC) imposed an embargo on oil shipments to the United States. Representative of the latter group are John C. Fisher's *Energy Crisis in Perspective* (New York: John Wiley & Sons, 1974) and Don E. Kash and Robert W. Rycroft's *U.S. Energy Policy: Crisis and Complacency* (Norman: University of Oklahoma Press, 1984). Robert Stobaugh and Daniel Yergin, eds., *A Time to Choose: America's Energy Future* (Cambridge, MA: Ballinger Publishing Co., 1974) concluded from an examination of the crisis that the United States had two options: either continue to depend on imported foreign oil or promote energy efficiency mainly through conservation while developing renewable energy sources. They did not see continued reliance on imports as a rational option for meeting the nation's energy needs. Yergin's Pulitzer Prize–winning *The Prize: The Epic Quest for Oil, Money, and Power* (New York: Touchstone Books: Simon & Schuster, 1992) sees the energy crisis as the result of a century-long trend in industrial states of mounting dependence on oil, and the consolidation of world production in the hands of countries who were eager to manipulate prices for national gain.

The energy crisis of the 1970s highlighted fundamental changes in American society in the twentieth century. The extravagant consumption of cheap energy had created a lifestyle of comfort and convenience, which appeared to be jeopardized by rising costs and dwindling supplies of energy. This is the theme of David E. Nye's *Consuming Power: Social History of American Energies* (Cambridge, MA: MIT Press, 1998). The universal adoption of the automobile spawned a car culture, a theme found in most histories of motor transportation in the United States. That culture is discussed in John B. Rae's *The American Automobile: A Brief History* (Chicago: University of Chicago Press, 1965); David L. Lewis and Lawrence Goldstein, eds., *The Automobile and American Culture* (Ann Arbor: University of Michigan Press, 1965); James J. Flink's *The Car Culture* (Cambridge, MA: MIT Press, 1988); and Jan Jennings, ed., *Roadside America: The Automobile in Design and Culture* (Ames: Iowa State University Press, 1990). The rising level of energy consumption in modern homes is one of the themes in Merritt Ierley's *The Comforts of Home: The American House and the Evolution of Modern Convenience* (New York: Clarkson Potter Publishers, 1999) and Jane Furnival's *Suck, Don't Blow: The Gripping Story of the Vacuum

Cleaner and Other Labour Saving Machines Around the House (London: Michael O'Mara Books Limited, 1998).

Government Publications

Sources ideal for tracking changes in the production and consumption of coal, oil, natural gas, and oil from shale are the publications of the U.S. Geological Survey and Bureau of Mines. Especially valuable for following the growth of Colorado's fossil fuel industries are the annual publications *Mineral Resources of the United States*, issued by the USGS from 1882 to 1923 and by the USBM from 1924 to 1931. Commencing in 1932–1933, the USBM changed the name to *Minerals Year Book*. Until 1977, each annual issue included sections devoted to coal, oil, natural gas, and, when appropriate, oil from shale in the states and the nation. A separate volume, titled "Area Reports, Domestic," incorporated into the *Minerals Year Book* from 1958 to 1993, provided data on minerals activities in each of the fifty states. "Area Reports; International" were issued from 1963 to 1988. Effective in 1978, the collection and publication of data about fuels were transferred from the Bureau of Mines to the new Department of Energy.

The USGS and the USBM also published various studies—*Bulletins, Professional Papers, Technical Papers, Information Circulars*, and *Reports of Investigations*—devoted to the discovery, development, production, and consumption of fossil fuels. *Bulletins*, which focused on geology, frequently included historical summaries of fuel production and distribution at many locations in Colorado. Particularly useful for the study of coal mining and coke-making are G. B. Richardson, "The Trinidad Coal Fields, Colorado," *USGS Bulletin 381* (1910): 379–446; Chester W. Washburne, "The Cañon City Coal Fields, Colorado," *USGS Bulletin 381* (1910): 341–378; and Paul Averitt, "Coking Coal Deposits in the Western United States," *USGS Bulletin 1222* (1966).

Equally valuable for the origin and evolution of Colorado's petroleum industry are N. M. Fenneman, "Structure of the Boulder Oil Field, Colorado," *USGS Bulletin 225* (1903): 383–391; "The Florence, Colo., Oil Field," *USGS Bulletin 260* (1905): 36–440; "Geology of the Boulder District, Colorado," *USGS Bulletin 265* (1905): 1–101; and Chester W. Washburne, "Development of the Boulder Field, Colorado," *USGS Bulletin 381* (1910): 514–516 and "The Florence Oil Field, Colorado," *USGS Bulletin 381* (1910): 517–544.

Indispensable for the study of the origins of the oil-from-shale industry in western Colorado are E. G. Woodruff and David T. Day, "Oil Shale of Northwestern Colorado and Northeastern Utah," *USGS Bulletin 581* (1915) and Dean E. Winchester, "Oil Shale in Northwestern Colorado and Adjacent Areas," *USGS Bulletin 641* (1917): 139–198. Government-sponsored efforts to advance technology for extracting oil from shale are discussed in Martin J. Gavin and John S. Desmong's "Construction and Operation of the Bureau of Mines Experimental Oil Shale Plant, 1925–1927," *USBM Bulletin 315* (1930). A synthetic-fuels program launched during World War II spawned another oil-from-shale boom. It is the subject of "The United States Bureau of Mines Synthetic Liquid Fuels Program, 1944–1955," *USBM Report of Investigations 5506* (1959). The principal technical achievement of the second round of public-funded research is reported in A. Matzic, J. R. Ruark, J. E. Phillips, J. D. Lankford, and Boyd Guthrie, "Development of the Bureau of Mines Gas-Combustion Oil-Shale Retorting Process," *USBM Bulletin 635* (1966).

On May 26, 1978, President Jimmy Carter established by executive order a commission to conduct a comprehensive review of coal mining and to give special attention to labor-management relations, living conditions in mining camps, and the effects of government regulations on the production and utilization of the solid fossil fuel. The commission prepared four publications during its tenure from 1978 to 1980, which were issued by the Congressional Office of Technology Assessment: "The American Coal Miner: A Report on Community and Living Conditions in the Coalfields," "The Direct Use of Coal: Prospects and Problems of Production and Combustion," "Coal Data Book," and "Recommendations and Summary Findings."

The United States Department of Energy, with exclusive authority to collect and publish data about fuels, placed responsibility for that function in a subordinate agency, the Energy Information Administration (EIA). Since 1978, the EIA has issued a variety of reports on the production, distribution, and consumption of fossil fuels. It has published, in addition, reports on the threats posed by the combustion of fossil fuels to the environment, and on the utilization of renewable energy sources.

The EIA's reports on coal allow researchers to follow the recent growth of the industry in Colorado and in the United States. Especially useful are *Two Decades of Change: Coal Production, 1970–1990* (Washington, DC, 1992), and *Coal Data: A Reference* (Washington, DC, 1995).

Essential for tracking petroleum production and consumption are the EIA's *Petroleum Supply Annual, 1982–1999* (Washington, DC, 1985–2000); *Petroleum 1996: Issues and Trends* (Washington, DC, 1997); and *The U.S. Petroleum Industry: Past as Prologue, 1970–1992* (Washington, DC, 1993). Statistical data for natural gas are in the EIA's *Gas Annual, 1993–1999* (Washington, DC, 1993–2000) and *Natural Gas: Issues and Trends 1998* (Washington, DC, 1999).

The EIA's *State Energy Data Reports: Consumption Estimates, 1883–1997* (Washington, DC, 1984–1999) are especially valuable for tracking changes in patterns of fuel consumption. These volumes contain, for Colorado and other states, consumption estimates for coal, natural gas, petroleum (with separate statistics for gasoline, fuel oil, jet fuel, etc.), nuclear electric power, hydroelectric power, and biomass. Similar data are provided for the state's residential, commercial, industrial, and transportation energy consumption, with estimates of energy input by electrical utilities.

Fossil fuels have accounted not only for most domestic energy production, but also for most of the man-made pollution in the form of carbon dioxide and other gases. Useful for tracking pollution are the EIA's *Emissions of Greenhouse Gases in the United States 1998* (Washington, DC, 1999) and *National Air Pollution Trends Update, 1970–1997* (Washington, DC, 1970–1997). The growing levels of pollution have focused attention in recent years on the need for nonpolluting alternative fuels. Their adoption has been limited, but significant, as shown in the EIA's *Renewable Energy Annual 1995–1998* (Washington, DC, 1996–1999).

Colorado's governmental agencies published reports and other materials related to the production and consumption of fossil fuels. With respect to coal, the most important are the Coal Mine Inspector's *Reports, 1884–1981* (Denver, 1885–1982). Issued biennially from 1885 to 1912, and annually from 1913 to 1981, each report contains statistics for injuries and fatalities in mines; the names of all active mines, identified by location and owner; a description of mining methods; statistics of employment and production; and recommendations for strengthening regulatory activities. Since 1982, the Department of Natural Resources, Division of Mines, State Coal Mine Inspection, has issued monthly summaries of coal-mining activities, including names and locations of properties, employment and production data, and on-the-job injuries and fatalities.

Fossil-fuels production, along with statistical data for consumption, appeared in the *Colorado Year Book*, published sometimes annually, at other times biennially, even sporadically, from 1918 to 1964. The State Board of Immigration issued the volumes from 1918 to 1933–1934. The State Planning Division continued the publication until 1964. These volumes are particularly useful for following year by year the rising output of natural gas in Colorado and its displacement of coal as a fuel of preference by householders, commercial interests, and industries.

The Colorado Geological Survey issued reports that related to the mining and utilization of fossil fuels in the state. Especially useful for following recent advances in output and consumption are *Colorado Minerals and Mineral Fuel Activities, 1994–1999* (Denver, 1995–2000). The Colorado Geological Survey joined the Colorado Oil and Gas Association and the Colorado Oil and Gas Conservation Commission in publishing *The Roads Less Traveled Make the Difference: A Colorado Perspective on Natural Gas and Oil in the Rockies* (Denver, 1998).

Valuable tools for research in the history of oil-from-shale programs in western Colorado are the Oil Shale Symposia, sponsored by the Colorado School of Mines from 1964 to 1992. The published proceedings of the twenty-five symposia contain commentaries on almost every aspect of oil-shale research and applications.

Trade Publications

Periodicals, most of which can be called trade publications, are major sources of information about fossil-fuels production, distribution, and consumption. The *Engineering and Mining Journal* (New York, 1886–1948) and *Mining and Scientific Press* (San Francisco, 1899–1921) focused on all aspects of mining. Each publication, in addition to articles on mining operations technology, production, distribution, and consumption of fuels, printed communications from correspondents in major mining camps on activities of interest to readers. *Mining and Scientific Press* merged with *Engineering and Mining Journal*, which was later absorbed by *Coal Age*.

Historically, three fossil fuels accounted for most energy production. Coal accounted for the largest share until 1951, and after a long interruption, again in the 1980s and 1990s. The issues of *Coal Age* (1909–1988) trace fluctuations in the industry's fortunes, with attention to mining enterprises, production, marketing trends, technology, and consumption. *Coal Age* was absorbed by *Coal* (1988–1996), which was succeeded by a new publication, *Coal Age* (1996 to the present).

Serving the coal-mining industry are the *Bituminous Coal Facts and Figures* (1948) and its successor, the *Bituminous Coal Annual* (1949–1953), published by the Bituminous Coal Institute, the public-relations arm of the National Coal Association. Each annual volume contains information about mining and consumption. In addition to statistical data on almost every aspect of the industry, there was a heavy emphasis on new technology, which had the potential of reducing costs and expanding markets.

The *Oil and Gas Journal* (1902 to the present) and the *Gas Journal* (1874 to the present) are devoted largely to the concerns of the American petroleum industry, with some attention to activities abroad. A leading regional trade magazine, published under different names, was the *Rocky Mountain Oil Reporter* (1947–1965), *Pacific States and Rocky Mountain Oil Reporter* (1966), and *Western Oil Reporter* (1966–1972). Whatever the title, it was a major source of information about oil and gas exploration and development in Colorado, and of the market forces that shaped production and consumption.

Newspapers

Information about fossil-fuels industries may be gleaned from Colorado's many newspapers. They should be used with care, because so-called news was sometimes nothing more than hearsay. However, in the absence of other sources, they may be all that is available for determining activities in some locations. Newspapers usually dwell on the personal side of coal, oil, and gas operations, giving attention to workers and their families and the conditions under which they worked and lived. Denver newspapers, because of their statewide news-gathering services, consistently carried a large volume of news about fuel mining and marketing activities. The *Rocky Mountain News* (1860 to the present), the oldest of the dailies; the *Denver Republican* (1881–1913); and the *Denver Post* (1892 to the present) are excellent sources for the study of fossil fuels. Elsewhere in the state, the value of newspapers may depend upon the tempo of fuel activities in specific localities. The *Boulder County Herald* (1880–1914) and the *Daily Camera* (1891 to the present) contain much information about the discovery, development, and decline of the Boulder oil field. The *Florence Refiner* (weekly, 1887–1901), initially called the *Florence Oil Refiner*, and the *Florence Citizen* (weekly, 1898–1914) are important sources for day-to-day activities in the state's first commercial oil field. The *Trinidad Daily*

Advertiser (1883–1898, 1901–1918), the (Pueblo) *Daily Chieftain* (1898–1973), and the (Pueblo) *Star-Journal* (1901–1973) have excellent coverage of fuel activities in southern Colorado.

Index

Page numbers in italics indicate illustrations.